Sustainability of Concrete With Synthetic and Recycled Aggregates

Edited by Hosam M. Saleh

Published in London, United Kingdom

IntechOpen

Supporting open minds since 2005

Sustainability of Concrete With Synthetic and Recycled Aggregates
http://dx.doi.org/10.5772/intechopen.94633
Edited by Hosam M. Saleh

Contributors
Amin Kamal Akhnoukh, Abebe Demissew Gashahun, Anatoly Krishan, Mauricio Pradena, Andrés César, Jose Bogas, Ana Carriço, Sofia Real, Mhammed Abdeldjalil, Rahmouni Abdelkader, Abdelghani Brahimi, Mourad Meghachou, Hicham Abbad, Redouane Chebout, Khaldoun Bachari, Mohammed Belbachir, Fatima Zohra Zeggai, Mohammed A. Abed, György L. Balázs, Hosam Saleh, Mohanad S. Eid

First published in London, United Kingdom, 2022 by IntechOpen
IntechOpen is the global imprint of INTECHOPEN LIMITED, registered in England and Wales, registration number: 11086078, 5 Princes Gate Court, London, SW7 2QJ, United Kingdom
Printed in Croatia

British Library Cataloguing-in-Publication Data
A catalogue record for this book is available from the British Library

Additional hard and PDF copies can be obtained from orders@intechopen.com

Sustainability of Concrete With Synthetic and Recycled Aggregates
Edited by Hosam M. Saleh
p. cm.
Print ISBN 978-1-83881-956-9
Online ISBN 978-1-83881-957-6
eBook (PDF) ISBN 978-1-83881-958-3

For EU product safety concerns:
IN TECH d.o.o., Prolaz Marije Krucifikse Kozulić 3, 51000 Rijeka, Croatia, info@intechopen.com or visit our website at intechopen.com.

Meet the editor

Hosam M. Saleh is a Professor of Radioactive Waste Management at the Radioisotope Department, Atomic Energy Authority, Egypt. He obtained an MSc and Ph.D. in Physical Chemistry from Cairo University. Dr. Saleh has more than twenty-five years of experience in hazardous waste management with an emphasis on treatment and developing new matrixes for the immobilization of these wastes. He is also interested in studying innovative economic and environmentally friendly techniques for managing hazardous and radioactive wastes. He has authored many peer-reviewed scientific papers and chapters and served as an editor of several books. He has been selected among the top 2% of scientists in the world according to a 2020 Stanford University report.

Contents

Preface

This book provides guidance on the treatment of environmental contamination arising from the cement industry and the use of concrete, a problem that requires the synergistic action of the scientific community.

The development and application of approaches and technologies that provide economic and sustainable development of concrete materials is an essential issue in concrete durability, structural design, and alternatives to traditional aggregates used in concrete material.

Authors of this book's chapters present advances in relevant fields related to assessing the sustainability of concrete with synthetic and recycled aggregate. The book contains nine chapters, organized into two sections that cover important research aspects of sustainable concrete technologies. The first section includes six chapters and opens with an introductory chapter that presents a brief overview of the characterizations of cement and modern sustainable concrete. Chapter 2 deals with structural and chemical analysis of new cement-based on eggshells and sand dunes (in southwest Algeria) stabilized with polyethylene terephthalate (PET). Chapter 3 discusses thermoactivated recycled cement, and Chapter 4 investigates the analysis of different approaches to sustainable concrete material. Chapter 5 examines the bearing capacity of concrete-filled steel tube columns, and Chapter 6 evaluates the freeze/thaw resistance of concrete.

The second section includes three chapters that discuss the production and applications of supplementary concrete. Chapter 7 discusses replacing cementitious materials for sustainable concrete productions, Chapter 8 discusses the application of supplementary cementitious materials in the precast concrete industry, and Chapter 9 discusses the application of a granular model to identify the granulometry of concrete material based on sand dunes.

The editor wishes to thank all the chapter authors for their valuable contributions and Author Service Manager Ms. Zvjezdana Tintor and the staff at IntechOpen for their assistance in finalizing this work.

Hosam M. Saleh
Egyptian Atomic Energy Authority,
Cairo, Egypt

Section 1

Analysis and Characterization of Modified Concrete

Chapter 1

Characterizations of Cement and Modern Sustainable Concrete Incorporating Different Waste Additives

Mohanad S. Eid and Hosam M. Saleh

Abstract

This chapter provides a brief introduction to cement and concrete, from their first utilization in constructions and monuments in different civilizations to their development and use in modern times. It discusses the modern forms of these materials, their physical and chemical properties, and their various applications. The chapter also examines the sustainability of concrete mixed with different waste additives, which can impart desirable properties to concrete, as a promising way to reduce environmental hazards resulting from the landfilling of these wastes. In addition to environmental benefits, waste utilization has economic benefits as well.

Keywords: cement, sustainable concrete, waste additives

1. Introduction

According to the *Encyclopedia of Science and Technology*, a hydraulic cementing agent (typically Portland cement), aggregate, water, and frequently regulated proportions of entrained air combine to form a flexible engineering material called "concrete." Concrete starts out as a flexible, workable slurry that can be shaped into a variety of forms. The hydration process between cement and water produces strength. The byproducts, which consist mostly of calcium silicates, calcium aluminates, and calcium hydroxide, are relatively insoluble and bond the aggregate in a cemented matrix [1].

Concrete is popular due to three distinguishing characteristics: plasticity, durability, and economy. When wet, concrete may be poured into almost any form, fit into almost any space, fill almost any vacuum, and coat nearly any surface. However, once it has dried and cured, it retains its shape, growing stronger, harder, and more settled over time [2].

Concrete may be waterproof, stormproof, and fireproof if it is built with the appropriate concentration and under the right circumstances. And because of its endurance, it may be expected to survive indefinitely. When all the steel we used to create our planet has rusted through and the wood has rotted into dust a million years from now, only concrete will remain.

Concrete is not a new or recent material. Like many other methods of building and development, it evolved over time. Our human ancestors discovered naturally existing materials that they could utilize to enhance essential components of their infrastructure, including houses, fences, wells, and so on. The generations that came after built on that knowledge, making small improvements until the advent of the Industrial Revolution, which accelerated building development to its current level [3].

The first concrete-like constructions were established in approximately 6500 BC by the Nabataean merchants, or "Bedouins" as they were called, who inhabited and ruled several oases and established a small empire between the north of Jordan and the south of Syria. They eventually discovered the benefits of hydraulic lime or cement that can gain strength and harden under water. Around 800 BC, they used kilns in the production of mortar to construct homes with rubble walls, floors using concrete, and subterranean watertight cisterns.

The Nabataeans knew the need of keeping the mixture in a dry or low-slump condition while creating concrete, as too much water produces cavities and flaws. Their construction methods included tamp down the fresh produced concrete using a special tool. This tamping process creates an additional gel, which is formed by chemical interactions that occur during hydration and bonds together the particles and aggregate [4].

More than 5000 years ago, its believed that the Egyptians used an early form of concrete to build pyramids. The blocks were limestone blocks hauled from quarries nearby. To hold those blocks together, the builders combined straw with mud consisting of crushed limestone, gypsum, and clay [3].

Around the same period, people in the north of China utilized a sort of cement in the construction of boats and the "Great Wall." Testing revealed that glutenous, sticky rice was a significant element in the mortar utilized to construct the Great Wall and many other ancient Chinese monuments. Some of these monuments have survived for a very long time and even recent demolition attempts [3, 5].

The Greeks found a natural pozzolanic mineral having hydraulic characteristics when combined with limestone around 600 BC, but they were not creative in using concrete like the Romans. After about 400 years, the Romans effectively employed concrete in their construction, but it was not the same as the concrete utilize today. It was more like solidified debris rather than a plastic, flowing substance poured into shapes.

The Romans built most of their constructions by piling up stones of varying sizes and filling in the gaps between the stones with mortar by hand. Above the ground, the walls were coated with bricks made of clay on both the interior and exterior, which also functioned as a form of concrete. The brick was structurally insignificant and was used mostly for decoration. Prior to this period, and in most areas at the time, mortars were made of a slowly hardening lime cement as it reacted with carbon dioxide from the air. These mortars were not effective because no obvious chemical hydration occurred [6].

It wasn't until the mid-18th century, when John Smeaton discovered a more contemporary technique of manufacturing hydraulic lime for cement, that technology advanced significantly. He utilized limestone that included clay and was burned until it became what is called a "clinker," which was then crushed into powder. This material was utilized in the reconstruction of the historical Eddystone Lighthouse in Cornwall, England [7].

Portland cement was developed in 1824 by Joseph Aspdin, who burned finely powdered chalk and clay until the carbon dioxide was eliminated. The cement was named after high-quality construction stones mined from Portland, England [3, 8].

Industrial structures improved greatly in the 19th century. Francois Coignet, who inserted rods made of steel into concrete to prevent the outside walls of construction to spread out, was the first to utilize iron-reinforced Portland cement concrete in houses in France and England.

2. Cementation material

Cement is a fine and dry powdery substance with major ingredients of calcined lime and clay, the former provides calcium oxide, and the later provides silica, alumina, and iron oxide.

When cement is mixed with water, it hardens into a solid mass. Setting and hardening are caused by hydration, which is a chemical reaction that combines the cement compounds with water to produce microscopic-level crystals or a gel-like material (calcium silicate hydrate) with a high specific surface area. The gel has adhesive properties as well as cohesive forces that can bind other solid materials together [9].

3. Cement, mortar, and concrete

The terms cement, concrete, and mortar are often used synonymously even though they are not. Cement, concrete, and mortar are actually three distinct materials [10].

Cement: As mentioned previously, cement turns into a gel-like material when mixed with water, having glue characteristics. Cement is never utilized alone, but it is an essential component of both concrete and mortar, binding their main components together.

Mortar: Mortar is a mixture of cement and fine aggregates (usually sand or other similar components) combined with water. It has the proper consistency to provide a thick, uniform lining that adheres securely to surfaces to bind materials together, for example, bricks, concrete blocks, or stones.

Concrete: Like mortar, concrete is composed of cement and fine aggregates mixed with water, but also coarse aggregates like graves, which makes it stronger and more durable.

Concrete has a lower water-to-cement ratio and thinner consistency than mortar, as well as more strength duo to the presence of the large aggregates.

There are three main types of concrete based on density.

Ordinary concrete: This is the most often utilized kind of concrete. It is employed in building construction where very high tensile strength is not required. It has moderate density around 2200–2500 kg/m^3.

Lightweight concrete: The single most essential characteristic of lightweight concrete is its extremely low heat conductivity. It is used for thermal insulation, construction blocks, and to protect steel buildings. It has low density less than 1900 kg/m^3 [11].

High-density concrete: This type of concrete is also known as heavy weight concrete. High-density materials are used as fine aggregates, coarse aggregates, or both. It is typically seen in nuclear power reactors and other comparable buildings because it offers excellent protection against all types of radiation. This concrete may have very high density of 3000–4000 kg/m^3 [12].

4. Properties and types of cement

4.1 Physical properties

Cement used are characteristic by its physical properties, which can control the quality of the cement. Some parameters of good cement are [13]:

1. fineness
2. soundness
3. consistency
4. setting time
5. strength
6. heat of hydration
7. loss of ignition

4.1.1 Fineness of the cement

Fineness is defined as the particle size of the cement. It is an important factor influencing the cement's rate of hydration. The reaction between cement and water occurs only at the surface of the solid particles, thus the accumulation of large particles on the surface of unreacted material could hinder the overall reaction.

The required fineness of good cement is achieved during the cement manufacturing process by grinding the clinker, with the most common method used for both controlling the grinding process and testing the finished cement [14].

One method is to measure the surface area per unit weight of cement by determining the rate of passage of air through a bed of cement [15].

4.1.2 Soundness of the cement

Soundness is defined as the immunity of cement to shrinking upon hardening. After the cement has set, it must not undergo any noticeable expansion, which can be caused by the presence of too much hard-burned free lime and magnesia. After setting, a good cement retains its volume without any deferred expansion.

Unsoundness in cement may appear after several years, so tests must be performed to ensure soundness and determine the possibility of this risk occurring [16].

- Le-Chatelier test

This method examines the expansion of cement caused by undesired excess lime. Cement paste is sandwiched between two glass plates and immersed in water for 24 hours at room temperature. It is then removed to measure the distance between the indicators before being returned to the water and boiled for at least an hour. The space between indicator points is measured again after the device has cooled. This distance should not be greater than 10 mm in a good-quality cement [17].

- Autoclave test

In this test, the cement paste is placed in an autoclave device (high-pressure steam vessel) to be cured. The pressure is slowly brought to about 2 MPa, and the temperature increased to 210°C. The past is then kept in the autoclave for three hours. After bringing the autoclave to room temperature and pressure gradually, the specimen's length change is measured and expressed as a percentage. A maximum

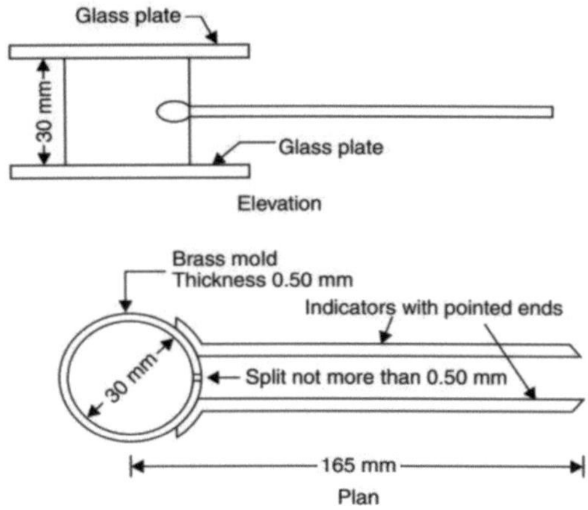

Figure 1.
Le-Chatelier apparatus [18].

of 0.8 percent autoclave expansion is required for good-quality cement according to the American Society for Testing and Materials (ASTM) (**Figure 1**) [17].

4.1.3 Consistency of cement

Consistence is the minimum amount of water required to complete the chemical reaction between water and the cementation material.

Determining standard consistency of cement is essential because if less water is added than the standard consistency would not complete chemical reaction and more water would increase water-cement ratio in either two cases will result in the reduction in strength. A Vicat apparatus is used to obtain the correct consistency of cement.

- Vicat test

The Vicat apparatus is loaded with cement paste and the plunger of the apparatus is lowered until it touches the top surface of the cement. The plunger, which has a length of about 50 mm and a diameter of about 10 mm, will penetrate the cement to a certain depth depending on its consistency. When the needle penetrates 10 mm into a cement, it is said that the cement has a normal consistency.

To obtain a standard or normal consistency of cement, only 25–35% water by weight of cement should be added (**Figure 2**) [19, 20].

4.1.4 Setting time of the cement

When water is added to cement, it hardens and sets. This period can vary according to several factors, including cement fineness, chemical content, cement–water ratio, and admixtures. Setting time could also be used to predict hydration rate. There are two important setting times that can be measured:

Initial setting time is the time it takes for the cement paste to lose its plasticity as soon as the water is added to cement.

Final setting time is the time taken for the cement to harden and completely lose its plasticity and be able to sustain some small loads.

A Vicat test can also be used to measure and test the setting time of the cement according to ASTM guidelines [20].

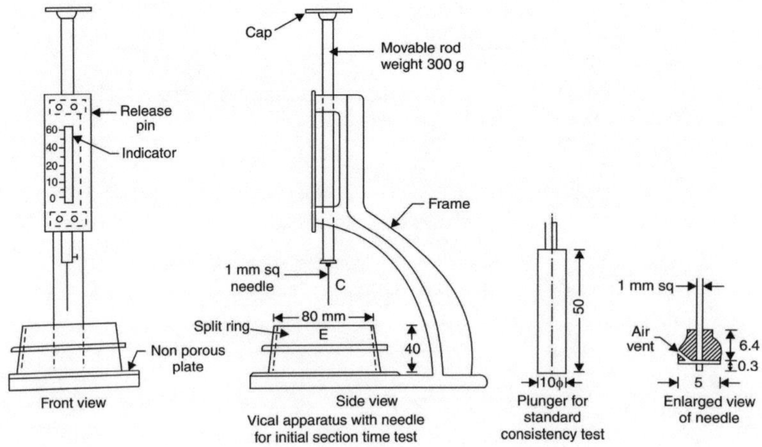

Figure 2.
Vicat apparatus [18].

4.1.5 Strength of the cement

Tests measuring the rate at which a cement can develop strength are usually carried out on mortar or concrete. There are numerous factors affecting strength, such as cement–fine aggregate ratio, water–cement ratio, the manner of mixing and molding, size and shape of a specimen, curing conditions, and age and loading conditions.

Before testing strength, it is important to note that cement strengthens over time; thus, a strength test should be performed at a specific time. Most cement strength tests are conducted at 3, 7, and 28 days. However, a one-day strength test may be performed for rapid hardening cement types.

- Compressive strength

In a test to determine compressive strength, a cylinder- or cube-shaped test specimen is subjected to a continuous compressive load under a hydraulic presser until failure. The loading sequence should be between 20 seconds and 80 seconds [21].

- Flexural strength

Flexural strength is a measurement of tensile strength in bending. The test is performed on a 40 x 40 x 160 mm rectangular concrete or mortar, which is subjected to a load at its center point until failure, according to ASTM [22].

4.1.6 Hydration heat of the cement

Hydration of cement generates heat, which can affect the cement's quality. When the heat generated is high, it may cause undesired stress, especially in large structures. The heat of hydration is affected mostly by the presence of tricalcium aluminate (C_3A) and tricalcium silicate (C_3S) in cement, in addition to fineness of the cement, water–cement ratio, and curing temperature. The heat of hydration of cement can be calculated according to ASTM [23]. This test method provides the apparatus and procedure for using isothermal conduction calorimetry to determine the total heat of hydration of cementitious materials at test ages up to 7 days.

4.1.7 Loss on ignition of the cement

Loss on ignition is a test used in inorganic analytical chemistry, especially for mineral analysis. It entails strong heating (igniting) of a sample of the material to a specific temperature while letting volatile chemicals escape until the mass of the sample reaches a constant value. In most cement types, water is lost around 100–105°C, organic material is burned at around 550°C, and most carbonates are lost between 800°C and 1000°C, as mentioned in ASTM [24].

4.2 Chemical properties

As mentioned previously, the main raw materials that make up cement are limestone and clay (silica, alumina, and iron oxide), as well as many other components including shells, chalk, and blast furnace slag. Chemical examination of cement raw materials reveals important information about cement's chemical properties [25].

Tricalcium silicate (C_3S), ($3CaO \cdot SiO_2$): This is an important compound that provides early strength of cement paste during hydration (initial sitting).

Dicalcium silicate (C_2S), ($2CaO \cdot SiO_2$): In contrast to tricalcium silicate, this compound enhances strength acquired after one week.

Tricalcium aluminate (C3A), ($3CaO \cdot Al_2O_3$): This compound is characteristic for its fast reacting with water, which causes an immediate stiffness of the cement paste. In addition, small amounts of C3A make cement resistant to sulfate.

Magnesia (MgO): A large amount of magnesia in cement can make it unsound and expand, but a small amount can make it stronger. CO_2 emissions are also minimized while MgO-based cement is produced. The MgO percentage of all cement is capped at 6% [26].

Tetracalcium aluminoferrite (C4AF), ($4CaO \cdot Al_2O_3Fe_2O_3$): Ferrite is a fluxing agent that lowers the raw material melting temperature in a kiln from around 1650–1420°C. Even though it hydrates quickly, it does not make a significant contribution to the hydration process [27].

Sulfur trioxide and free lime: Just like magnesia, excessive amounts can cause expansion and the cement to become unsound.

Alkalis: The alkali content of cement is determined by the quantities of potassium oxide (K_2O) and sodium oxide (Na_2O). Too much alkali in cement might make it difficult to control the cement's setting time. When used with calcium chloride in concrete, low-alkali cement can induce discoloration [28].

Alumina: Because alumina is chemical-resistant, cement with a high alumina content may tolerate frigid temperatures. It also speeds up the setting time but weakens the cement [29].

Silica fume: This is used to increase a range of qualities in cement, including compressive strength, abrasion resistance, and bond strength. Although the use of silica fume extends setting time, it can provide extremely high strength. As a result, cement with a silica fume content of 5–20% is typically used in high-strength cement projects [30].

5. Different types of cement

Cements with different chemical compositions may have various properties. As a result, it should be possible to select a mixture of raw materials to produce a cement with desired properties.

In general, cement is divided into two kinds based on the method of hardening and setting:

- **Hydraulic cement**: Cements that harden when exposed to water, also produce water-resistant products. The main raw materials used to synthesize hydraulic cement are limestone, clay, and gypsum. Hydraulic cement is made by burning this raw material at a very high temperature, according to ASTM.

- **Non-hydraulic cement:** Water is not required for the hardening of non-hydraulic cement because it absorbs carbon dioxide (CO_2) from the atmosphere. To harden, this sort of cement requires dry circumstances. Non-hydraulic cement involves lime, gypsum plasters, and oxychloride as basic materials.

5.1 Portland cement

Portland cement is the most common, low-cost, and widely used type of cement worldwide because its basic ingredients are inexpensive and readily available. It consists of four major components: tricalcium silicate, tricalcium aluminate, dicalcium silicate, and tetracalcium aluminoferrite.

For specialized objectives, such as durability and high early strength, several kinds of Portland cement are produced to fit distinct physical and chemical characteristics [31, 32].

- **Ordinary Portland cement (Type I):** This all-purpose cement may be used in any situation where specific characteristics are not required.

- **Modified Portland cement (Type II):** This is a cement that is commonly utilized in situations with abnormal sulphate concentrations in groundwater and soils. It is also used when a cement with a moderate heat of hydration is required. The concentration of the C3A shall not exceed 8%.

- **Rapid hardening cement (Type III):** This cement attains high strength in several days. The chemical makeup of this cement is the same as that of Type I Portland cement, but it has greater C3S content with finer ground. Its strength after 24 hours is approximately comparable to that of conventional Portland cement after 3 days with the same water ratio. The use of this cement results in significant time and cost savings.

- **Low-heat cement (Type IV):** This cement is used when the level and rate of heat generation must be kept to a minimum. The percentages of C2S and C4AF are quite high, whereas the percentages of C3S and C3A are comparatively low. A small bit of tricalcium aluminate (usually less than 7%) causes the concrete to have a low heat of hydration. This type of cement is ideal for mass concrete structures such as dams.

- **Sulphate resisting cement (Type V):** This is a cement that is commonly used in concrete subjected to severe sulphate conditions in harsh environments, particularly when soils or groundwaters have a high sulphate concentration. Type V cement builds strength slower than Type I cement does, due to high silicate content with low C3A and C4AF concentrations.

- **Air-entraining cement (Type IA, Type IIA, and Type IIIA):** These cement types have the same properties as Type I, Type II, and Type III cements. Grinding a tiny quantity, approximately 0.05%, of animal and vegetable fats, oil, and another acid with a wetting agent such as aluminum powder, hydrogen peroxide, and so on results in the entrainment of extremely fine air bubbles in the concrete, which improves workability with a lower water–cement ratio, as well as improves frost-resistant characteristics.

- **Hydrophobic cement:** Hydrophobic cement is made from regular Portland cement clinker by grinding it with specific water repellent components such as oleic acid, stearic acid, or naphthenic acid. A water-resistant coating is formed over each cement particle, preventing water or moisture from the air from being absorbed by the cement. This film is broken during the regular hydration process and behaves just like ordinary Portland cement.
  This cement has excellent resistance during transportation and long-term storage in severely moist climatic conditions.

- **Quick-setting cement:** This is cement that has the property of setting very quickly. This feature is obtained by decreasing the gypsum content during the clinker grinding process.
  Quick-setting cement is used for concreting in static or flowing water, underwater construction, and rainy and cold weather conditions, or when work must be finished in a short period of time, like in places with higher temperatures where water evaporates easily.

The distinction between rapid hardening and quick-setting cement is that quick-setting cement sets faster. Simultaneously, the rate of strength growth is comparable to that of ordinary Portland cement, but quick hardening cement builds strength rapidly.

5.2 Pozzolanic cement

Pozzolanic cements are blends of Portland cement with pozzolanic material, which might be natural or synthetic. Natural pozzolanas are mostly volcanic in origin like diatomaceous earth. Fly ash, burnt clays, and shales are examples of materials utilized in synthetic pozzolanic.

Pozzolanas are materials (natural or synthetic) that have no cementitious properties, but they contain reactive silica (and alumina). When divided into fine form, they are capable of combining with lime (calcium hydroxide) in the presence of water to create compounds having cementitious characteristics [2, 32].

This cement is widely utilized because it has a strong resistance to different chemical attacks when compared to regular Portland cement.

5.3 High alumina cement

High alumina cement is a rapid hardening cement produced by fusing a combination of bauxite (aluminum ore) and limestone in a reverberatory or electric furnace or rotary kiln at 1500–1600°C.

The cement consists of about 30–40% lime, 45–50% alumina, up to 10% iron oxides, and preferably no more than 6% silica. Calcium aluminate ($CaO \cdot Al_2O_3$) is the main cementing ingredient.

High alumina cement has a high early strength, as it can reach ultimate strength within 24 hours, a high heat of hydration, and a very high durability against high temperatures, frost, and chemical attacks [2, 33].

5.4 Slag cements

Slag is the glass-like byproduct left over after separating a desired metal from its raw ore, and it is generally generated via a blast furnace–oxygen converter method or electric arc furnace. The primary components of these slags are calcium, iron, silicon, magnesium, and aluminum oxides, with lower quantities of phosphorus, manganese, and others depending on the raw materials utilized [2, 34].

- **Blast-furnace slag cement:** This is a mixture obtained by adding Portland cement or grinding the clinkers with about 65% granulated slag. It can be utilized for projects when economic considerations are important.

- **Super-sulphated cement:** This is a slag-containing cement with less than 6% sulfuric anhydride (SO_3), at least 8% granular slag, 10–20% hard-burned gypsum or anhydrite (natural anhydrous calcium sulphate), and a few percentages of Portland cement. This cement is used in severe situations such as maritime construction, mass concrete projects to withstand the aggression of strong seas, and chemical works exposed to high concentrations of sulphates.

6. Sustainable concrete

Mortar and concrete can have various properties and applications according to the type of cement used and the other components of fine and large aggregates.

The future of concrete research and engineering is trending towards employing waste materials. These wastes, which can be hazardous and difficult to dispose of, may be present in large quantities in the environment [35].

Employing waste in concrete can be achieved by adding it or using it to replace one of the other components, which can achieve or even improve desired properties.

7. Metallic waste additives

The amount of industrial metallic waste is huge and can be challenging to recycle. These metals, however, can be added to concrete and mortar to improve their properties. In this context, Małek et al. utilized metal chip waste generated by lathes and CNC machines as an additive to concrete. Results show that adding metal waste can improve the physical and mechanical properties of concrete [36].

Norambuena-Contreras et al. utilized two kinds of steel industrial wastes, steel shavings and steel wool fibers, to manufacture mortar. They concluded that there were small variations in flexural and compressive strength, but an overall improvement of the mortar due to adding the steel [37].

8. Polymeric waste additives

Polymers have been increasingly used over the last 50 years in almost every industry as well as in human daily life. However, these products are hazardous to the environment and almost impossible to recycle. Therefore, finding a useful way to reuse these products is a focus of research. One potential way to reuse polymers is to add them to concrete or mortar.

Daud et al. utilized thermoplastic polyethylene terephthalate (PET) waste obtained from recycling bottles to replace the graves in lightweight concrete used for insulation walls in building construction faces [38].

Plastic pollution is exacerbated by the constant manufacture and disposal of plastic. Jain et al. successively employed non-metallic polymeric plastic waste with flay ash in concrete and concluded that the addition of non-metalized waste plastic bag fibers together with fly ash in concrete greatly improved properties like split tensile strength, resistance to abrasion, flexural strength, impact, and drying shrinkage [39].

Another type of sustainable concrete incorporates waste rubber as an additive. Rubber is one of the hardest materials to recycle, thus the utilization and reuse of leftover waste rubber in concrete can minimize raw material consumption, resulting in economic efficiency and long-term development of the building sector. Yang et al. utilized waste tire rubber as a replacement for fine aggregates (sand) by reducing its size. The disposal of leftover waste tires has become a major environmental concern worldwide [40].

Concrete containing waste polymeric material can be utilized to reduce the contamination of radioactive material in the environment. Saleh et al. incorporated recycled PET waste with cement into making container-like material for reducing the activity of radioactive borate waste obtained from pressurized water reactor (PWR). They provided a characterization for these samples and concluded that the combination of cement with PET is an acceptable product for immobilization of radioactive borate [41].

9. Biowaste additives

Biomass and agriculture waste can be great additives or even replacements to cement in concrete and mortar. Because cement is a key building material that accounts for 8–10% of total CO_2, there is a critical need to replace cement in order to reduce greenhouse gases (GHGs) such as CO_2, which are responsible for global warming [42].

The research of agricultural wastes as extra ingredients in mortar and concrete, as well as replacements for aggregates, resulted from the quest for alternative materials for cement. Agricultural wastes are readily available, inexpensive, and have high reactivity in concrete, which has increased researchers' interest in these materials. The effective use of these elements in concrete will result in a green manner of disposing of enormous amounts of trash that might otherwise have caused environmental harm.

Bassam et al. provide an excellent review on previous studies that use rice husk waste as a replacement for cement. The influence of rice husk on concrete characteristics such as workability, flexural strength, splitting tensile strength, density, compressive strength, modulus of elasticity, durability abilities, and sustainability is reported [43].

Not only the biomaterial itself but also its byproduct ashes can be used for making a sustainable cementitious material.

Blessen et al. provide a comprehensive review based on previous research and current developments in concrete using biomass ashes from agricultural farming wastes, ashes from bamboo leaves, date palm, rice straw, olive waste, elephant leaves, banana leaf and plantain peels, rice straw, wheat straw, olive waste, and corn cob as pozzolanic materials in cement concrete [44].

Some plants can be harmful and dangerous to the environment like the aquatic plant *Myriophyllum spicatum*, which is an invasive and noxious. Saleh, et al.

incorporated this plant into cement for safe immobilization of radioactive isotopes and other hazardous waste. Results show the cementation of dry solid waste generated during phyto-remediation was effective in immobilizing these wastes [45].

10. Glass waste additives

Waste glass is another readily accessible resource that may be utilized in place of cement. Only a portion of waste glass is reused in the production of new glass, and the remainder is discarded owing to impurities, color, or cost.

Glasses are categorized into 32 different kinds, although the most common are soda-lime, vitreous silica, borosilicate, lead, aluminosilicate, alkali silicates, barium, and aluminosilicate glasses [46].

Crushed waste glass has a variety of characteristics, including a high concentration of silicon and calcium and an amorphous structure. As a result, waste glass possesses pozzolanic or cementitious characteristics and can be utilized as a partial substitute for cement.

Ankur et al. published a review of studies that employed silica fume and waste glass into concrete as replacements for cement and provided results on physical and mechanical properties [47].

Author details

Mohanad S. Eid[1] and Hosam M. Saleh[2*]

1 Faculty of Science, Physics Department, Tanta University, Tanta, Egypt

2 Nuclear Research Center, Radioisotope Department, Egyptian Atomic Energy Authority (EAEA), Egypt

*Address all correspondence to: hosamsaleh70@yahoo.com

References

[1] Trefil J. The Encyclopedia of Science and Technology. New York: Routledge; 2001

[2] Shetty MS, Jain AK. Concrete Technology (Theory and Practice). 8e ed. New Delhi: S. Chand Publishing; 2019

[3] Jahren P, Sui T. History of Concrete: A Very Old and Modern Material. Singapore: World Scientific; 2017

[4] Baker I. Concrete. In: Fifty Materials That Make the World. Switzerland: Springer; 2018. pp. 35-42

[5] Richard WS. History of Concrete, PDF. Aberdeen Group. Arch. from Orig., Vol. 28. 2015

[6] Brandon CJ, Hohlfelder RL, Jackson MD, Oleson JP. Building for Eternity: The History and Technology of Roman Concrete Engineering in the Sea. United Kingdom: Oxbow Books; 2014

[7] Bensted J, Coleman N. Cement and concrete-7000 BC to 1900 AD. Cement, Wapno, Beton. 2003;**8**(no. 70, nr 3): 134-142

[8] Herring B, Miller S. The Secrets of Roman Concrete. Constructor, Virginia: Associated General Contractors of America (AGC); 2002

[9] Scrivener KL, Nonat A. Hydration of cementitious materials, present and future. Cement and Concrete Research. 2011;**41**(7):651-665

[10] Li VC, Maalej M. Toughening in cement based composites. Part I: Cement, mortar, and concrete. Cement and Concrete Composites. 1996;**18**(4): 223-237

[11] Zhang MH, Gjvorv OE. Mechanical properties of high-strength lightweight concrete. Materials Journal. 1991;**88**(3): 240-247

[12] Mortazavi SMJ, Mosleh SMA, Maheri MR, Yousef NIAH, Zoulghadri S and Hajipour A. Production of an economic high-density concrete for shielding megavoltage radiotherapy rooms and nuclear reactors. International Journal of Radiation Research. 2007;**5**(3):143-146.

[13] Powers TC, Brownyard TL. Studies of the physical properties of hardened Portland cement paste. Journal Proceedings. 1946;**43**(9):101-132

[14] Bentz DP, Sant G, Weiss J. Early-age properties of cement-based materials. I: Influence of cement fineness. Journal of Materials in Civil Engineering. 2008; **20**(7):502-508

[15] Goodwin RW. Combustion Ash Residue Management: An Engineering Perspective. Kidlington, UK: William Andrew; 2013

[16] Kabir H. Evaluation of the Autoclave Expansion Test for Cement. Canada: University of Toronto; 2019

[17] Kabir H, Hooton RD, Popoff NJ. Evaluation of cement soundness using the ASTM C151 autoclave expansion test. Cement and Concrete Research. 2020;**136**:106159

[18] Bhavikatti SS. Basic Civil Engineering, New Age, 2010: Basic Civil Engineering. New Delhi: New Age International (P) Limited; Vol. 1. Bukupedia; 2010

[19] CEN E. 196-3 Methods of testing cement-Part 3: Determination of setting times and soundness. Brussels, Belgium: European Committee for Standardization; 2017.

[20] C. ASTM. Standard Test Methods for Time of Setting of Hydraulic Cement by Vicat Needle. 2008 C191-08.

[21] Norma A. C109/C109M, Standard Test Method for Compressive Strength of Hydraulic Cement Mortars (Using 2-in. or [50-mm] Cube Specimens). West Conshohocken, Pa. EE. UU: American Society for Testing and Materials; 1999

[22] A. C348. Standard Test Method for Flexural Strength of Hydraulic-Cement Mortars. West Conshohocken, Pennsylvania, United States: American Society for Testing and Materials; 2008

[23] A. C1702-17. Standard Test Method for Measurement of Heat of Hydration of Hydraulic Cementitious Materials Using Isothermal Conduction Calorimetry. 2017

[24] ASTM C114 - 18. Standard Test Methods for Chemical Analysis of Hydraulic Cement. West Conshohocken, Pennsylvania, United States: American Society for Testing and Materials; 2018

[25] Abd El-Hafiz NA, Abd El-Moghny MW, El-Desoky HM, Afifi AA. Characterization and technological behavior of basalt raw materials for Portland cement clinker production. International Journal of Innovative Science Engineering and Technology (IJISET). 2015;**2**(7)

[26] Bao Y, Zhu Y, Zhong W, Qian F. A novel chemical composition estimation model for cement raw material blending process. Chinese Journal of Chemical Engineering. 2019;**27**(11):2734-2741

[27] Carlson ET. Some Properties of the Calcium Aluminoferrite Hydrates. Vol. 6. US Government Printing Office; 1966

[28] Jiang S, Kim B-G, Aïtcin P-C. Importance of adequate soluble alkali content to ensure cement/ superplasticizer compatibility. Cement and Concrete Research. Gaithersburg, US: National Institute of Standards and Technology; 1999;**29**(1):71-78

[29] Midgley HG, Midgley A. The conversion of high alumina cement. Magazine of Concrete Research. 1975; **27**(91):59-77

[30] Rao GA. Investigations on the performance of silica fume-incorporated cement pastes and mortars. Cement and Concrete Research. 2003;**33**(11):1765-1770

[31] Levy SM. "Calculations relating to concrete and masonry," Construction Calculation Manual. MA, USA: Butterworth-Heinemann; 2012. pp. 211-264

[32] ASTM C150 / C150M-21. Standard specification for Portland cement. ASTM International. Waltham, USA: Butterworth-Heinemann; 2021. DOI: 10.1520/C0150_C0150M-21

[33] Zieliński K, Kierzek D. The impact of alumina cement on properties of portland cement slurries and mortars. International Journal of Structural Engineering. 2021;**15**(6):323-326

[34] Liu Y, Zhang Z, Hou G, Yan P. Preparation of sustainable and green cement-based composite binders with high-volume steel slag powder and ultrafine blast furnace slag powder. Journal of Cleaner Production. 2021; **289**:125133

[35] Collivignarelli MC, Abbà A, Miino MC, Cillari G, Ricciardi P. A review on alternative binders, admixtures and water for the production of sustainable concrete. Journal of Cleaner Production. 2021; **295**:126408

[36] Małek M, Kadela M, Terpiłowski M, Szewczyk T, Łasica W, Muzolf P. Effect of metal lathe waste addition on the mechanical and thermal properties of concrete. Materials (Basel). 2021; **14**(11):2760

[37] Kanellopoulos A. J. N.-C. and A. C. and I. G.-T. and M. C. Effect of metallic

waste addition on the physical and mechanical properties of cement-based mortars. Applied Sciences. 2018;**8**(6): 929. DOI: 10.3390/app8060929

[38] Daud MAM, Selamat MZ, Rivai A. Effect pf thermoplastic polymer waste (PET) in lightweight concrete. In: 2nd International Conference on Sustainable Materials (ICoSM 2013). Vol. 795. 2013. pp. 324-328. DOI: 10.4028/www.scientific.net/AMR.795.324

[39] Jain A, Sharma N, Choudhary R, Gupta R, Chaudhary S. Utilization of non-metalized plastic bag fibers along with fly ash in concrete. Construction and Building Materials. 2021;**291**:123329

[40] Li Y, Zhang S, Wang R, Dang F. Potential use of waste tire rubber as aggregate in cement concrete – A comprehensive review. Construction and Building Materials. 2019;**225**:1183-1201. DOI: 10.1016/j.conbuildmat.2019.07.198

[41] Saleh HM, Tawfik ME, Bayoumi TA. Chemical stability of seven years aged cement–PET composite waste form containing radioactive borate waste simulates. Journal of Nuclear Materials. 2011;**411**(1-3):185-192

[42] Raheem AA, Ikotun BD. Incorporation of agricultural residues as partial substitution for cement in concrete and mortar – A review. The Journal of Building Engineering. 2020;**31**:101428. DOI: 10.1016/j.jobe.2020.101428

[43] Tayeh BA, Alyousef R, Alabduljabbar H, Alaskar A. Recycling of rice husk waste for a sustainable concrete: A critical review. Journal of Cleaner Production. 2021;**312**:127734

[44] Thomas BS, Yang J, Mo KH, Abdalla JA, Hawileh RA, Ariyachandra E. Biomass ashes from agricultural wastes as supplementary cementitious materials or aggregate replacement in cement/geopolymer concrete: A comprehensive review. The Journal of Building Engineering. 2021;**40**:102332

[45] Saleh HM, Moussa HR, El-Saied FA, Dawoud M, Bayoumi TA, Wahed RSA. Mechanical and physicochemical evaluation of solidified dried submerged plants subjected to extreme climatic conditions to achieve an optimum waste containment. Progress in Nuclear Energy. 2020;**122**:103285

[46] Chen G et al. Glass recycling in cement production—an innovative approach. Waste Management. 2002;**22**(7):747-753. DOI: 10.1016/S0956-053X(02)00047-8

[47] Mehta A, Ashish DK. Silica fume and waste glass in cement concrete production: A review. The Journal of Building Engineering. 2020;**29**:100888. DOI: 10.1016/j.jobe.2019.100888

Chapter 2

Structural and Chemical Analysis of New Cement Based on Eggshells and Sand from Dunes (Southern West of Algeria) Stabilized by PET

Abdelghani Brahimi, Mourad Meghachou, Hicham Abbad, Abdelkader Rahmouni, Redouane Chebout, Khaldoun Bachari, Fatima Zohra Zeggai and Mohammed Belbachir

Abstract

In this chapter, we present our study of geopolymers and hybrid geopolymers synthesized with treated fly ash from eggshells (FAES) and sand from the dunes of southern Algeria using activators such as NaOH and Na_2SiO_3, respectively, in addition to the organic polymer polyethylene terephthalate (PET). Several parameters have been modified, such as alkali concentration and percentage of activators and PET, with the objective to improve the quality of the desired geopolymers and hybrid geopolymers. The main objective of this work is to study the use of waste PET in the matrix of this new material to replace Portland cement, which is widely used today, as well as develop ecological building materials that are durable and lightweight and prevent chemicals from attacking old structures. Through optical and electron microscopy, we studied the effect of the addition of PET on the structure of our geopolymer material and on the bond and interface areas between the aggregates and the matrix. The microstructural analysis discussed here refers to specimens containing 5% PET by weight. We observed that PET contents significantly altered the structure and morphology of the samples.

Keywords: fly ash of eggshells (FAES), sand dune, cement, microstructure, analysis, construction, Young's modulus, structure, silica fume

1. Introduction

Geopolymers can replace cement in various forms of construction work and the manufacture of concrete and mortars, and they have very important mechanical characteristics given their three-dimensional aluminosilicate network [1]. The use of geopolymers instead of Portland cement is justified by the reduction in CO_2 emissions and energy savings [2]. Dune sand is a widely available natural resource that can be integrated into the construction industry, which needs more development [3]. The process of geopolymerization of mixtures of sand and fly ash is used to produce a new material composed of very fine elements. Mortars composed of modified polymers are building materials with excellent properties and can replace mortars based on Portland cement [4]. Polymers have been used to enhance the waterproof properties and modify the mechanical properties of concrete and mortars as well as reinforce adhesion [5]. The literature specifies that the characteristics of concrete and mortar modified by polymers depend mainly on the polymer percentage or on the ratio between the cement and polymer, that is, the ratio of the value of the mass of solid polymers contained in a polymer-based addition to the value of cement in a polymer, concrete, or modified mortar [6]. Among plastics, polyethylene terephthalate (PET) is the most widely used to produce products such as consumer goods, beverage bottles,and food packaging [7]. PET bottles have replaced traditional glass bottles for liquid storage due to their ease of handling, light weight, and possibility of storage [8]. Fly ash is a fine and powdery material produced from coal when generating electricity. When coal is used in a power plant, it is crushed into a very fine powder that will be blown into the furnace of the plant [9]. The hydrogen and carbon in the coal are depleted, leaving molten noncombustible particles rich in alumina and silica. After the solidification of these particles in the fly ash state, these very fine powders easily enter the atmosphere and pollute water and the air [10]; they can cause respiratory problems if not properly eliminated. In addition, fly ash deposited on leaves and plants in agricultural fields near electrical power plants can diminish crop yield. However, when used properly, fly ash can help to conserve natural resources [11]. The manufacture of Portland cement is a major contributor of CO_2 gas emissions, thus any reduction in the use of cements will result in a reduction in greenhouse gas emissions, possibly reducing emissions to zero. One ton of CO_2 is emitted for each ton of Portland cement produced [12]. Thus, replacing Portland cement with fly ash will eliminate CO_2 emissions. If all the fly ash produced is used instead of carrier cement in the various construction works of buildings, roads, and bridges, it is estimated that the reduction in CO_2 emissions will be equivalent to the elimination of 25% of vehicles worldwide [13]. This chapter summarizes the scientific advances in the preparation, fabrication, properties, and applications of fly ash of eggshells (FAES) and sand dune-based geopolymer and geopolymer hybrids. The production of mixed geopolymers and hybrid geopolymers is mainly based on alkali-activated geopolymerization, which can occur under mild conditions and is considered a cleaner process due to much lower CO_2 emissions than that from the production of cement [14].

2. Experiment

2.1 Materials and methods

2.1.1 Materials

The basic material used in these experiments is the original siliceous sand of the sand dunes of southwest Algeria. First, 100 g of dry sand was treated with 200 ml of hydrochloric acid over a period of 30 minutes at room temperature. Then, all the leaching tests were carried out in a 250-ml glass beaker placed on a magnetic stirrer with a control unit to ensure the homogeneity of the product at stable temperatures. When the required temperature (80°C) of the contents of the beaker (100 ml of acid) was reached, approximately 30 g of dry sand was added to the beaker while the contents of the beaker rotated at a constant speed of 250 rpm. The beaker was covered to avoid losses by evaporation. From the leaching solution, a sample amount of our prepared mixture was taken at predetermined time intervals, filtered, washed several times with distilled water to remove any unspent acid, and then dried at 110°C for 1 hour. All the experiments were repeated for more precision. The chemical composition of the silica sand prepared for this study was determined by X-ray fluorescence (XRF). **Table 1** presents the study results.

The eggshells were used as raw material to prepare the fly ash. They were washed first with distilled water and then with an acid solution (1 M HCl) to reduce the level of limine (CaO) and remove impurities. After drying, and when all moisture was removed, the material was calcined at temperatures ranging from 700–850°C after drying at 25°C. XRF analysis was performed to characterize the fly ash sample (see **Table 2**). It is evident that the sample is very rich in silica and eliminated, which makes it a suitable raw material to begin the geopolymerization process. The dimensions of the fly ash particles are less than 100 μm. The main element in pork roosters is CaO (63.69%).

Water glass, also called sodium silicate (Na_2SiO_3), was synthesized at the chemistry laboratory of Polymer Oran 1, University of Algeria. Then, 100 g of sand was washed with an acid solution (1 M HCl), dried at 25°C, and mixed with 200 g of sodium hydroxide (13 M NaOH). A platinum crucible was placed in an electric furnace at 850°C for 1 hour at a heating rate of 5°C/min. The mixture was melted, and the melt was cooled and solidified in the crucible. This method allowed us to synthesize 75 g of water glass nanomaterials (sodium silicate). This white powder was dried at 25°C to allow us to carry out microstructural, chemical, and mineralogical analyses.

Thermoplastic polyester (PET) has acceptable mechanical characteristics; in particular, a tensile modulus of elasticity of 2.89 GPa and a flexural modulus of elasticity of 2.36 Gpa, with a tensile strength of 58 Mpa and resistance to chemical attack. It is a semicrystalline polymer with a density melting point (specific gravity) of 1.28 to 1.39 g/cm3. The very fine grinding of plastic waste, such as drink bottles, gives us PET powder.

2.1.2 Methods

We analyzed the morphology of the raw fly ash samples using scanning electron microscopy (SEM;LEO SEM 1450), the molecular structure using Fourier-transform infrared spectroscopy (FTIR;Perkin Elmer 100 spectrum), the mineralogy using X-ray diffraction (XRD; X-ray generator, Philips PANalytical

Oxide	Content %	Oxide	Content %
SiO_2	90.83	Na_2O	0.01
Al2O3	7.45	CaO	1.02
Fe2O3	0.29	MgO	0.00
K_2O	0.01	TiO_2	0.09

Table 1.
Chemical composition of the sand dune (western Algeria).

Oxide	Content %	Oxide	Content %
SiO_2	0.06	Na_2O	2.92
Al2O3	0.04	CaO	63.69
Fe2O3	0.01	MgO	0.57
K_2O	0.04	TiO_2	0.02

Table 2.
Chemical composition of the fresh fly ash of eggshell (FAES).

pw3830), and the chemical composition using XRF (dispersion spectrometer, Philips 1404 wavelength).

2.2 Synthesis of different forms of geopolymers

The synthesis of geopolymers is carried out by mixing source materials containing aluminosilicate (sand dune), fly ash with a high calcium (Ca) classified as (F), and an alkaline solution (NaOH 13 M). The source materials used are FAES washed with acidic solution (HCl 1 M) to eliminate impurities and minimize the rate of calcium and limine as producers of carbon dioxide (CO_2), calcined at 700 to 900°C, and sand dune (Algerian sand) leached with acid solution (HCl 1 M) and mixed with alkaline solution (NaOH 13 M) to prepare sodium silicate (Na_2SiO_3). In the last step, four samples of geopolymers were prepared for comparison, as shown in **Table 3**. The final product was placed in molds at room temperature for 24 hours and then stored in an oven at 80°C for another 24 hours. **Table 4** presents the results of the XRF assay analysis. The sample is full of silica and limine, which is a suitable raw material to begin the geopolymerization process.

2.3 Synthesis of hybrid geopolymers

The synthesized organic–inorganic hybrid geopolymers consist of FAES and Na_2SiO_3 from sand dunes (southern Algeria) activated by alkaline solution (13 M NaOH) in which a percentage of PET was incorporated. Four different formulations of hybrids were prepared and characterized, as shown in **Table 5**.

FAES and sand from the dunes were used as the principal sources of aluminosilicates because they are the cheapest aluminosilicates with a good degree of purity. Moreover, this raw material improves the mechanical strength of and reduces salts and CO_2 in the final product. **Table 2** presents the compositions of the raw material. The mixture of fly ash and sand was sieved to obtain a fine powder with an average diameter of 100 μm. The powder was added gradually to an alkaline solution (NaOH 13 M) previously prepared by mixing a sodium silicate solution with SiO2/Na2O ratio = 2 and sodium hydroxide with Na2SiO3/NaOH ratio = 1.5. The final

Geopolymers (GP)	Mass (g) ratio
GP1: FAES, Na_22SiO_3, NaOH, H_2O	3.00, 1.75, 1.00, 2.25
GP2: FAES, Na_2SiO_3, NaOH, H_2O, SiO_2	3.00, 1.75, 1.00, 2.25, 0.5
GP3, FAES, Na_2SiO_3, NaOH, H_2O, Al_2O_3	3.00, 1.75, 1.00, 2.25, 0.5
GP4, FAES, Na_2SiO_3, NaOH, H_2O, Fe_2O_3	3.00, 1.75, 1.00, 2.25, 0.5

Table 3.
Geopolymers synthesis procedure (GP).

Oxide	Content %	Oxide	Content %
SiO_2	17.21	Na_2O	30.57
Al2O3	0.01	CaO	23.72
Fe2O3	0.11	MgO	0.22
K_2O	0.06	TiO_2	0.03

Table 4.
Chemical composition of the synthesized geopolymer (GP-2).

Geopolymers hybrid (GHPs)	Mass (g) ratio
GHP1: FAES, Na2SiO3, H2O, PET	3.00, 1.75, 1.00, 2.25, 0.5
GHP2: FAES, Na_2SiO_3, NaOH, H2O, SiO_2, PET	3.00, 1.75, 1.00, 2.25, 0.5, 0.5
GHP3: FAES, Na_2SiO_3, NaOH, H_2O, Al_2O_3, PET	3.00, 1.75, 1.00, 2.25, 0.5, 0.5
GHP4: FAES, Na_2SiO_3, NaOH, H_2O, Fe_2O_3, PET	3.00, 1.75, 1.00, 2.25, 0.5, 0.5

Table 5.
Geopolymers hybrid synthesis procedure (GHP).

composition of the synthesized geopolymer can be expressed as Si, Al, Na, and H_2O. To synthesize geopolymer/PET hybrid systems, PET was also added in part to the alkaline solution as well as to the mixture of fly ash and sand dune during the mixing phase. The resulting product was stirred mechanically for approximately 30 min to reach good homogenization and then poured into plexiglass-clad molds and sealed. The molds were placed in an oven for 24 hours at 30°C to avoid any possible thermal degradation of the polymer and successively stored for 28 days at room temperature.

3. Results and discussion

Geopolymers and hybrid geopolymers or cementitious materials in general were synthesized using FAES activated by alkali solution and silicate sodium from sand from the dunes in southern Algeria. Results confirm that quartz is the main component in the different forms of the prepared geopolymers, along with calcite, hematite, mullite, and ferrite. These are the main elements responsible for networking in geopolymer matrixes. Thus, reactivity under alkaline conditions is affected only in the amorphous section of these reagents and acts as an indicator of geopolymer and a substitute for metakaolin.

3.1 Analysis of geopolymers and hybrid geopolymers

3.1.1 Analysis of different form of geopolymers

The prepared geopolymers generally contain a large percentage of quartz in the form of silica (SiO_2), which exhibits better resistance to external actions due to the hardness of the material. Local conditions and sources influence the chemical composition of geopolymers [15]. XRF is the most reliable technique for finding the lowest concentrations of the elements in a prepared sample, as shown in **Table 4**.

Ceramic materials can be prepared by geopolymers. The preparation and synthesis of our geopolymers is carried out by a chemical solution of alkali silicate. Solid aluminosilicates have been added from the source of the sand of the dunes. FAES is very rich in calcium, and with the presence of 13 M NaOH as an alkaline solution, the XRD models studied show remarkable differences in the influence of the fly ash samples on the geopolymers due to the position and shape of the quartz peaks and bumps (lower and upper) (**Figure 1**). The XRD diagrams of the geopolymer based on fly ash confirm that the geopolymer materials are essentially composed of an amorphous character under X-ray, knowing that the diffraction crystals are the same as the original materials (calcite, mullite, limine, hematite, and quartz). An amorphous peak was observed, as the value of 2θ on the diffraction pattern ranged from approximately 20 degrees to 69 degrees, given the presence of amorphous glassy materials [16]. With activation of the ash by NaOH (alkaline

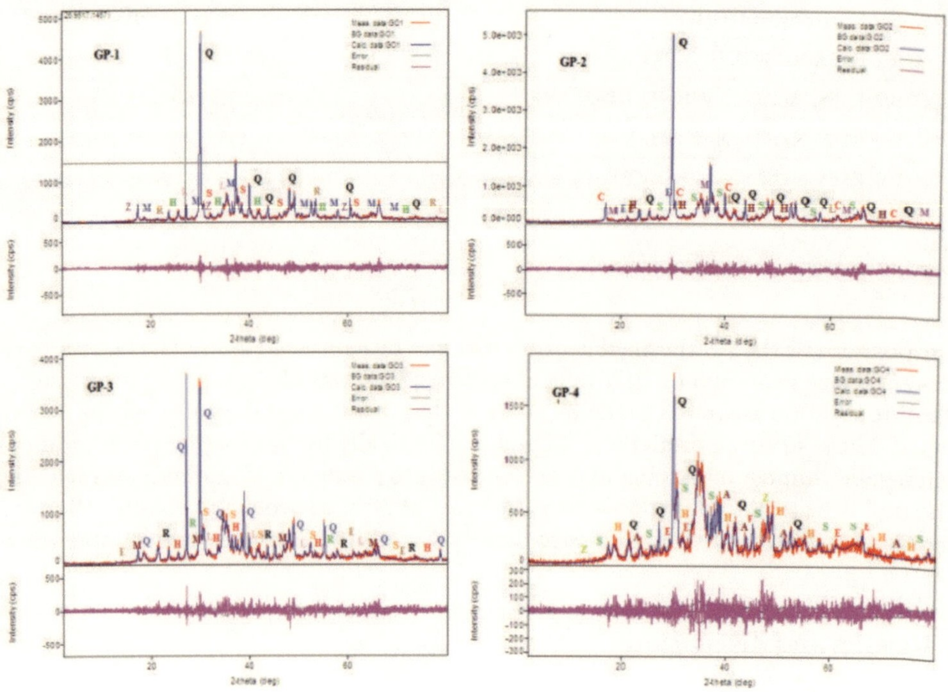

Figure 1.
XRD patterns of geopolymers (GP1, GP2, GP3, and GP4) with a NaOH molar ratio variation of 13 M.

solution), it was found that the diffractogram of the original ash was changed [17]. A slight shift of 19–50 degrees to 20–69 degrees (2θ) of the value of the peak attributed to the phase of the vitreous form of the original fly ash was observed. Ahydrate gel in the form of alkali aluminosilicate was formed as a result of this transformation. This hydrate gel is the main material that allows the initial reaction of geopolymerization of the geopolymeric materials mentioned in the diffraction diagrams [18]. With activation, the crystalline stages (hematite, quartz, calcite, mullite) observed inside the initial material remained practically without modification [19]. For the geopolymer model, the basic fly ash mineralogy does not change much, a result that agrees with the literature [20].

The SEM images shown in **Figures 2** and **3** show a change in morphology in most geopolymer samples compared to fly ash. Sodium silicate (Na_2SiO_3) was the most present element in all the samples studied (GP1, GP2, GP3, and GP4). In addition, after approximately 1 hour of preparation, a greater quantity of fly ash reacted positively. The microstructure of the geopolymers was heterogeneous and the matrix was full of loosely structured fly ash grains of different sizes, except in sample GP2, as shown in **Figure 3**, in which we observed a good microstructure. In the gel, several circular shaped cavities do not appear. Here, we suggest that a significant amount of spherically shaped fly ash reacts, and this result shows complete transformation within the system after only 1 hour and the reaction up to 78.93%. Finally, we found that the number of reactions taking place in a paste that forms the geopolymer develops as a function of the molar percentage of SiO_2/Al_2O_3 and the reactivity ratio of the fly ash, which is rich in calcium (Ca). The "GP" geopolymers prepared by us have many characteristics that make them economical materials for construction [21]. The formation of geopolymer-type concretes is accomplished by the addition of water to a geopolymer, keeping in mind that the different shapes of geopolymers have a relatively porous structure. The water molecules limit the formation of gases

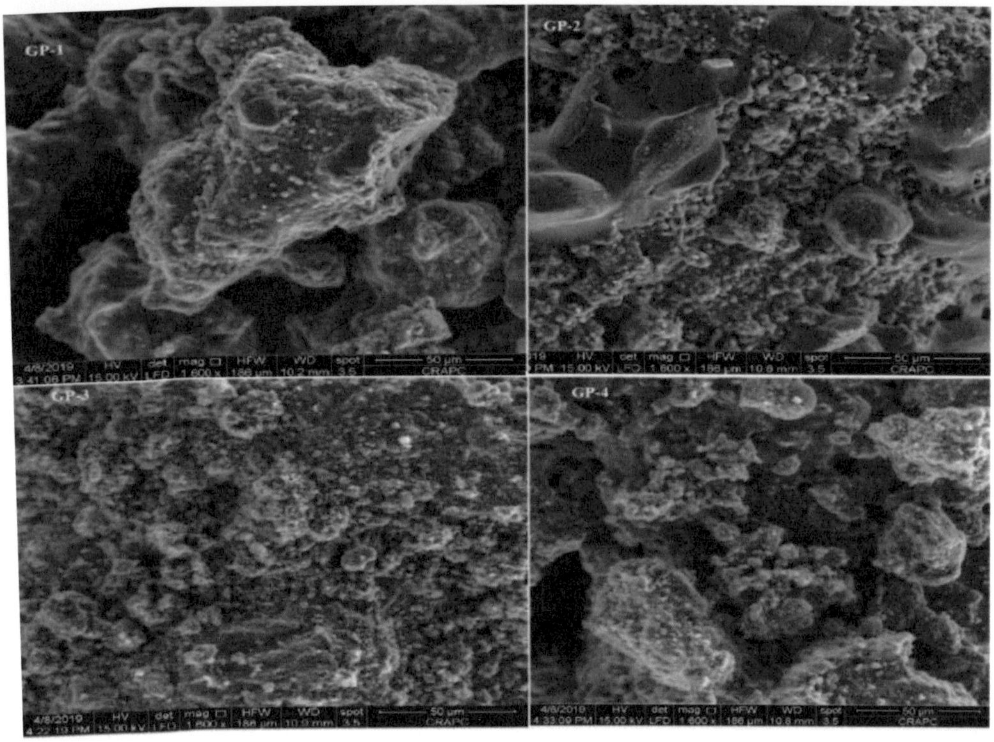

Figure 2.
SEM of the foamed geopolymer according to the percentage of fly ash eggshell after 1 hour (GP1, GP2, GP3, GP4).

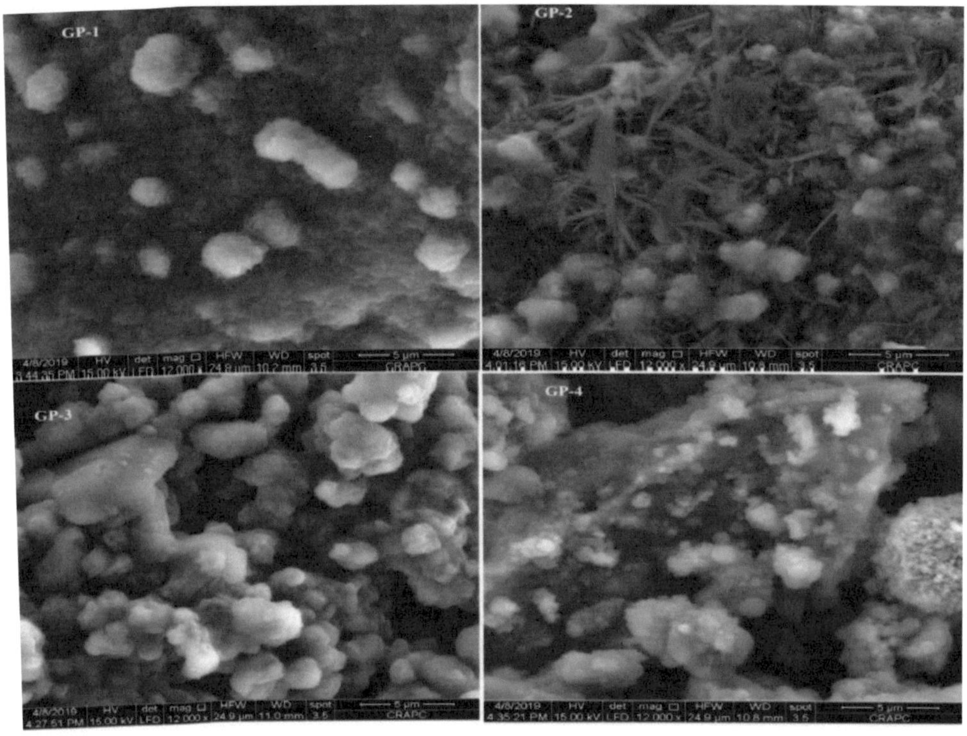

Figure 3.
SEM of the foamed geopolymer according to the percentage of fly ash eggshell after 24 hours (GP1, GP2, GP3, GP4).

before the gel hardens in the structure [22]. The geopolymer is formed when water is added to FAES in alkaline activation solutions. After adding FAES to the geopolymers, the number of gas-producing compounds is known, and they are trapped to produce a microstructure within the cured material [23]. When mixing water and geopolymers, the chemical reactions liberate different gases that are entrapped in the structure, especially CO_2. When the oxides of silicone and calcium as well as metallic aluminum in an alkaline solution are conserved, carbon oxide and H2 are removed, creating aluminum hydroxide. Finally, the CO2 molecules will be blocked in the structure of the geopolymer, which means that the product is very reactive.

3.1.2 Analysis of different forms of prepared geopolymer hybrids

XRD techniques were used for a wide variety of material characterization studies. The XRD results show some qualitative differences in the hydration rate due to the incorporation of PET polymer. **Figure 4** shows the X-ray patterns of the composites with 5% PET and composites without polymer PET (**Figures 3** and **4**). The main compounds observed are large amounts of $CaCO_3$, calcium oxide, and quartz resulting from anhydrous fly ash and sand dune, respectively. The peak intensity in 18, 52 degrees to 67, 37 degrees has been considered a region of the quantity of $CaCO_3$. Therefore, it is also noted here that at a hybrid geopolymer ratio of 5%, a slight increase in the peak intensity compared to the unmodified geopolymer is observed. Furthermore, $CaCO_3$ crystals may produce sharper reflection in the presence of PET polymer due to a change in the orientation pattern of the crystals. Based on SEM micrograph analysis (**Figures 5** and **6**) and chemical composition (**Table 6**), it is possible to explain this variation in terms of the fact that PET addition causes a progressive decrease in the amount of calcium carbonate and calcium

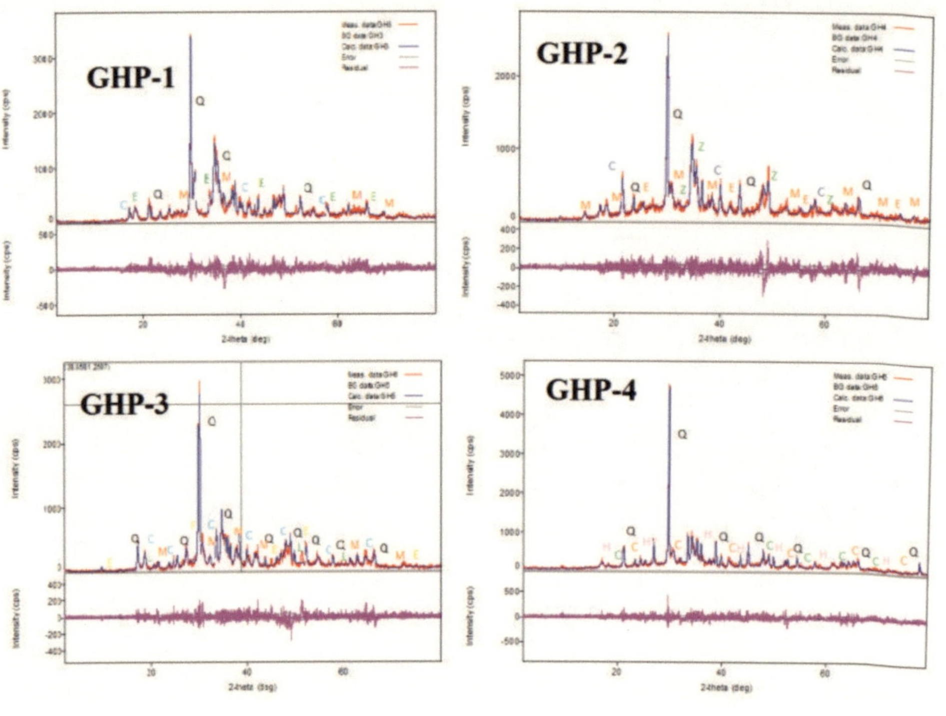

Figure 4.
XRD patterns of hybrid geopolymers (GHP-1, GHP-2, GHP-3, and GHP-4) with a NaOH molar ratio variation of 13 M.

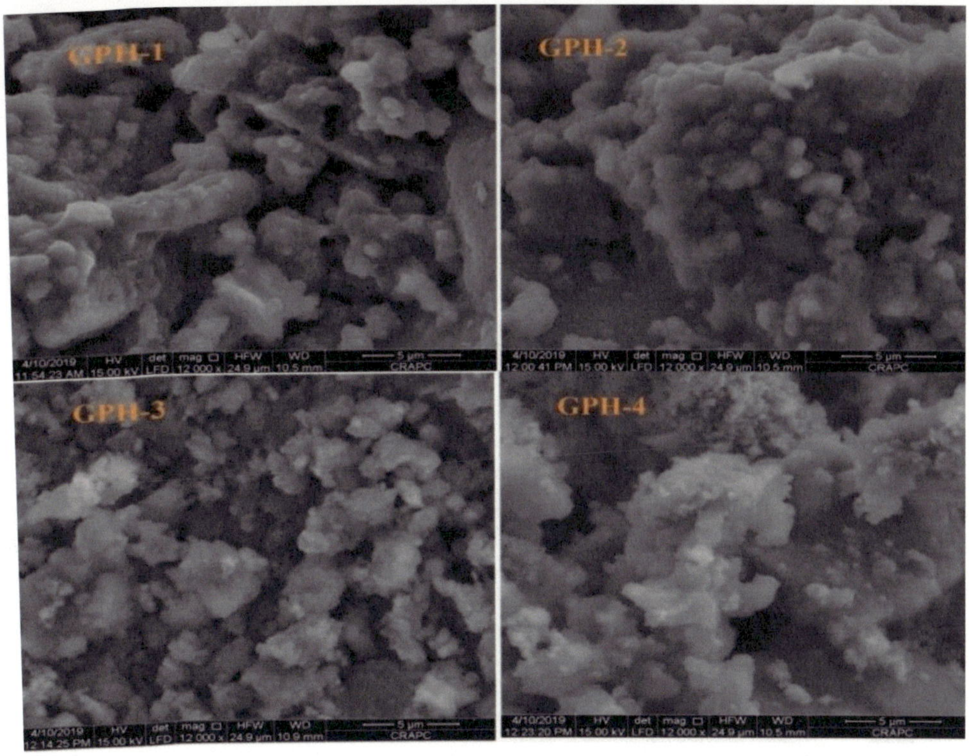

Figure 5.
SEM of the foamed hybrid geopolymer according to the percentage of fly ash eggshell after 1 hour (GPH1, GPH2, GPH3, GPH4).

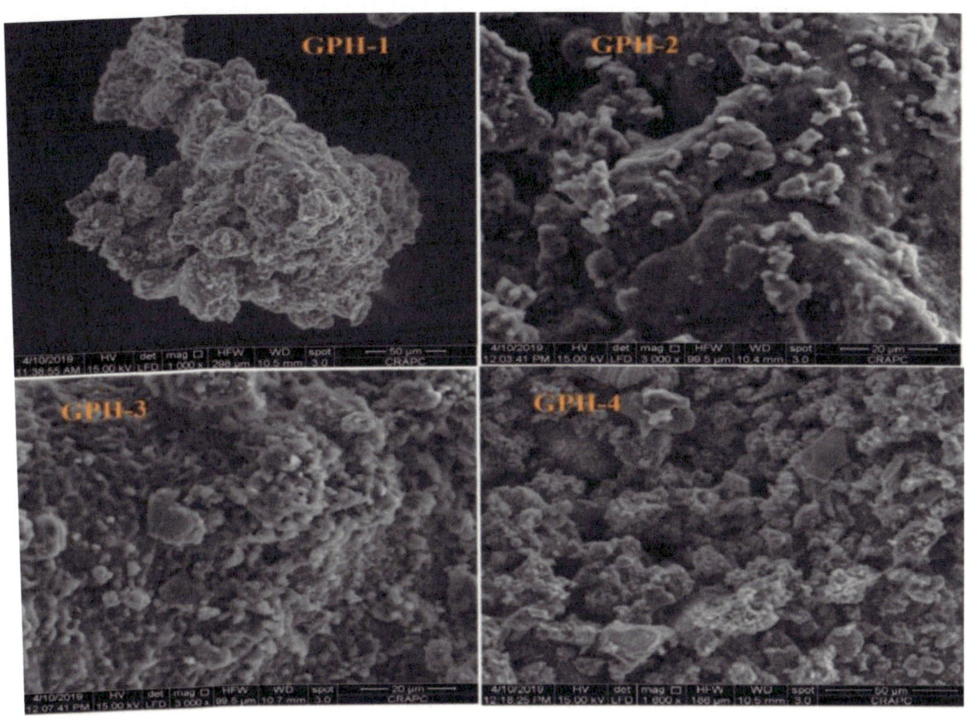

Figure 6.
SEM of the foamed hybrid geopolymer according to the percentage of fly ash eggshell after 24 hours (GPH1, GPH2, GPH3, GPH4).

Oxide	Content %	Oxide	Content %
SiO_2	14.22	Na_2O	23.11
Al2O3	13.42	CaO	19.78
Fe2O3	0.07	MgO	0.19
K_2O	0.04	TiO_2	0.02

Table 6.
Chemical composition of the synthesized hybrid geopolymer (GHP-2).

oxide in the hybrid composites compared to the unmodified geopolymer. These results agree with previous studies [24].

Images of the SEM micrograph in **Figures 5** and **6** show the microstructure of the mixture hybrid geopolymer based on PET with sand dune (quartz) after 1 hour. The PET composite indicates a weak interfacial transition zone in the hybrid geopolymer matrix interface at 1 hour, which is related to the large shrinkage of the geopolymer paste, as revealed by XRD [25, 26]. There is approximately less distance between PET and the surface of the geopolymer matrix. Furthermore, some microcracks also formed in the hybrid geopolymer. As shown in **Figures 5** and **6**, PET was placed in the crack zone when the specimen was subjected to a chemical reaction after 1 hour. The PET might deform and lengthen, rupture, or pull out because of the applied stress at the fiber section in a crack zone.

4. Conclusion

The synthesis of a new geopolymer-type material using a hydrothermal process was carried out after the preparation of base materials such as silica and alumina. This work describes the synthesis and valuation of FAES in the raw and activated states as well as sand from the dunes of southern Algeria for the preparation of geopolymer mortars by alkaline activation. The characterization and preparation of sodium silicate (Na_2SiO_3) were carried out in specialized laboratories in Algeria. Chemical analysis by XRF shows that the sand of the dunes is very rich in quartz ranging from 90.04% to 99.16% silica; in addition, a low concentration of other oxides was observed. However, microscopic SEM observations of the sand revealed the presence of pores with differing morphologies (e.g., rounded, elongated, and angular). The influence of FAES is important with respect to water penetration, and this performance is better than that of ordinary Portland cement. In addition, we observed very good fire resistance with a reduction in CO_2 emissions. SEM also revealed uniform and correct distribution of the eggshells over the whole matrix phase. It was also found that there is good adhesion between the sand particles and eggshell particles, due to the sand being rich in aluminum and silicon. From the alkaline activation of eggshells, it was found that FAES can be used with sand from dunes to obtain geopolymers and hybrid geopolymers that can be used as green and durable concrete.

Acknowledgements

We thank all staff of the research center in physicochemical analysis (CRAPC) of Tipaza, Algeria, for their kind cooperation and characterization (XRF, X-ray, SEM). We wish the best of luck to Prof. Mohammed Belbachir (Oran 1 University) and Prof. Mourad Megachou (University DjillaliLiabes of Sidi Bel Abbes) for their support in making the publication of this research possible.

Author details

Abdelghani Brahimi[1], Mourad Meghachou[1], Hicham Abbad[1], Abdelkader Rahmouni[2*], Redouane Chebout[3], Khaldoun Bachari[3], Fatima Zohra Zeggai[2,3] and Mohammed Belbachir[2]

1 Laboratory of Civil Engineering and Environment, Department of Civil Engineering, Djillali Liabes University, Sidi Bel-Abbes, Algeria

2 Laboratory of Polymer Chemistry (LCP), Department of Chemistry, University of Oran1 Ahmed Benbella, Oran, Algeria

3 Center for Scientific and Technical Research in Physico-chemical Analysis (CRAPC), Tipaza, Algeria

*Address all correspondence to: ramaek23@yahoo.fr

References

[1] Ganesan K, Rajagopal K, Thangavel K. Rice husk ash blended cement: Assessment of optimal level of replacement for strength and permeability properties of concrete. Construction and Building Materials. 2008;**22**:1675-1683

[2] Onera A, Akyuzb TS, Yildiza R. "An experimental study on strength development of concrete containing fly ash and optimum usage of fly ash in concrete". Cement and Concrete Research. 2005;**35**:1165-1171

[3] Sakulich AR. Reinforced geopolymer composites for enhanced material greenness and durability. Sustainable Cities and Society. 2011;**1**(4):195-210

[4] Duxson P, Fernández-Jiménez A, Provis JL, Lukey GC, Palomo A, van Deventer JSJ. Geopolymer technology: The current state of the art. Journal of Materials Science. 2007;**42**(9):2917-2933

[5] Ranjbar N, Talebian S, Mehrali M, Kuenzel C, Metselaar HSC, Jumaat MZ. Mechanisms of interfacial bond in steel and polypropylene fiber reinforced geopolymer composites. Composites Science and Technology. 2016;**122**:73-81

[6] Filho JH, Medeiros MHF, Pereir E, Helene P, Isaia GC. High volume Fly ash concrete with and without hydrated lime: Chloride diffusion coefficient from accelerated test. Journal of Materials in Civil Engineering. 2013;**25**:411-418

[7] Dr Soma NJ, Chandrasekhar D. A comparative study on egg Shell concrete with partial replacement of cement by Fly ash. International Journal for Research in Applied Science & Engineering Technology (IJRASET). 2015;**3**(Special Issue-I1)

[8] Chindaprasirt P, Rukzon S. "Strength, porosity and corrosion resistance of ternary bland Portland cement", rice husk and fly ash mortar. Construction and Building Materials. 2008;**22**:1601-1606

[9] Mathur VK, Verma CL, Gupta BS, Agarwal SK, Kumar A. "Use of Higher Volume Fly Ash in Concrete for Building Sector", Report No. T(S) 006. Roorkee: CII CANMET-CIDA, HVFA, Project, Environmental Science and Technology Division; 2005

[10] Kuenzel C, Vandeperre LJ, Donatello S, Boccaccini AR, Cheeseman C. Ambient TemperatureDrying shrinkage and cracking in metakaolin-based geopolymers. Journal of the American Ceramic Society. 2012;**95**(10):3270-3277

[11] Zuhua Z, Xiao Y, Huajun Z, Yue C. Role of water in the synthesis of calcined kaolin-based geopolymer. Applied clay. Science. 2009;**43**(2): 218-223

[12] Ridtirud C, Chindaprasirt P, Pimraksa K. Factors affecting the shrinkage of fly ash geopolymers. International Journal of Minerals, Metallurgy, and Materials. 2011;**18**(1):100-104

[13] Perera D, Uchida O, Vance E, Finnie K. Influence of curing schedule on the integrity of geopolymers. Journal of Materials Science. 2007;**42**(9): 3099-3106

[14] Glasby T, Day J, Genich R, Kemp M. Commercial scale geopolymer concrete construction. In: Proceedings of the Saudi International Building and Constructions Technology Conference. 2015

[15] van Riessen A, Chen-Tan N, Portella J, Bernard JS, Gourley T. Chapter 14: Geopolymer cement and concrete, 441 – 458. In: Ward C, Heidrich C, Yeatman O, editors. Coal

Combustion Products Handbook – Second Edition. Ash Development Association of Australia; 2014

[16] Williams R, van Riessen A. Determination of the reactive component of fly ashes for geopolymer production using XRF and XRD. Fuel. 2010;**89**:3683-3692

[17] Rickard WDA, Williams R, Jadambaa T, van Riessen A. Assessing the suitability of three australian fly ashes as an aluminosilicate source for geopolymers in high temperature applications. Materials Science and Engineering A. 2011;**528**:3390-3397

[18] Cement Industry Federation. Cementing Our Future 2005-2030. Technology Pathway for the Australian Cement Industry; 2005

[19] Mishra SB, Langwenya SP, Mamba BB, Balakrishnan M. Study on surface morphology and physicochemical properties of raw and activated south African coal and coal fly ash. Physics and Chemistry of the Earth. 2010;**35**:811-814

[20] Temuujin J, Williams RP, Riessen A. Effect of mechanical activation of fly ash on the properties of geopolymer cured at ambient temperature. Journal of Materials Processing Technology. 2009;**209**:5276-5280

[21] Sakorafa V, Michailidis K, Burragato F. Mineralogy, geochemistry and physical properties of fly ash from the megalopolis lignite fields, Peloponnese, southern Greece. Fuel. 1996;**75**:419-423

[22] Abdullah MMA, Jamaludin L, Hussin K, Bnhussain M, Ghazali CMR, Ahmad MI. Fly ash porous material using Geopolymerization process for high temperature exposure. International Journal of Molecular Sciences. 2012;**13**:4388-4395

[23] Álvarez-Ayuso E, Querol X, Plana F, Alastuey A, Moreno N, Izquierdo M, et al. Environmental, physical and structural characterization of geopolymer matrixes synthesised from coal (co-)combustion fly ashes. Journal of Hazardous Materials. 2008;**154**: 175-183

[24] Chindaprasirt P, Rattanasak U. Utilization of blended fluidized bed combustion (FBC) ash and pulverized coal combustion (PCC) fly ash in geopolymer. Waste Management. 2010;**30**:667-672

[25] Škvarla J, Sisol M, Botula J, Kolesárová M, Krinická I. The potential use of fly ash with a high content of unburned carbon in geopolymers. Acta Geodynamics et Geomaterialia. 2011;**162**:123-132

[26] Davidovits J. SPE PATEC. Brookfield Center, USA: Society of Plastic Engineering; 1979

Chapter 3

Thermoactivated Recycled Cement

José Alexandre Bogas, Ana Carriço and Sofia Real

Abstract

The cement industry is currently faced by the great challenge of reducing its vast carbon footprint, due to being the second highest industrial greenhouse gases (GHG) emitter. This value is expected to further increase, since cement production is foreseen to rise by about 20% until 2050. Therefore, more eco-efficient alternatives to ordinary Portland cement have been developed towards a sustainable concrete industry. This chapter presents some of the latest advances in low-carbon thermoactivated recycled cements (RC) obtained from old waste concrete, leading to a significant reduction of the GHG emissions, while also encouraging the valorization reuse of waste materials and the reduction of natural resource depletion. The manufacture and general performance of RC, including the main production issues, rehydration behavior and phase and microstructure development, as well as its incorporation in cement-based materials are discussed. Some of the most recent research, main challenges and future perspective of RC are addressed.

Keywords: recycled cement, thermoactivation, low-carbon binder, sustainable concrete, waste concrete

1. Introduction

The construction industry is one of the most relevant sectors contributing to global warming, involving the extensive consumption of raw materials, depletion of non-renewable resources, extensive greenhouse gas (GHG) emissions and significant construction and demolition waste (CDW) disposal. Therefore, stringent environmental measures and relevant international agreements have been established to reduce the environmental impact of this industry. In this context, large companies worldwide have been investing in more sustainable practices, promoting the development of alternative more eco-efficient building materials towards a truly circular economy supported on recycling, resource efficiency and low carbon emissions [1, 2]. Two major goals of the European council are the reduction of GHG emissions to 60–80% by 2050, on a 1990 base year and the reuse of at least 70% of CDW, excluding backfilling operations [3, 4].

In particular, concrete, as the most used building material in the world and involving substantial extraction of raw materials, significant GHG emissions and extensive CDW generation, becomes a serious source of environmental concern [5, 6]. Cement is the main concrete constituent responsible for the significant carbon footprint of concrete, accounting with over 80% of the total CO_2 emissions of concrete production [7]. In fact, over 5% of the world's anthropogenic CO_2 emissions are attributed to the cement industry [8, 9]. This alarming value is expected to further increase due to the ever-growing demand for this product, which is expected to rise by over 20% in 2050 [8, 9]. Therefore, the concrete industry and the scientific community have been focused on the urgent development of more sustainable eco-efficient concrete. The use of recycled aggregates in concrete production has been largely studied [10], but its effective acceptance in the construction sector is still a way off. On the one hand, the available technology is not efficient in providing high quality recycled aggregates with minor contamination with adhered paste that reduces their physical, mechanical and durability properties. On the other hand, the simple substitution of natural aggregates with recycled aggregates fails to significantly reduce the GHG emissions, which, as mentioned, are essentially related to cement production. Therefore, various strategies have been implemented concerning the efficient reduction of cement's environmental impact [8, 11, 12], namely the development of carbon capture solutions, new and more efficient production technologies, alternative fuels, alternative cements and the reduction of the clinker to cement ratio. Among these carbon reduction levers, the most promising and effective solution lies in carbon capture, but its viable implementation has yet to overcome some challenges [13]. The use of supplementary materials as partial clinker substitution has been considered for many years, but further GHG emissions reduction with currently available mineral additions is hard to explore. Moreover, the availability of some of these additions, namely those that are by-products of pollutant industries, such as fly ash, is becoming scarce.

Alternative low-carbon cements, such as calcium aluminate and alkali-activated cements have also been the object of intense research, but their implementation in the construction market is still far from being economically viable [13, 14]. More recently, a very promising approach relies on the production of low-carbon recycled cements (RC) from dehydrated waste hardened cement. The idea is to recover the binding properties of waste cement through its thermal activation at low temperatures, reducing the thermal energy of the clinker manufacture and avoiding the limestone decarbonation phase, which represents about 60% of the carbon emissions during the sintering process [13, 15].

The rehydration capacity of concrete subjected to high temperatures has long been shown from post-fire studies of concrete behavior [16–18]. This recovery was found to be related to the regeneration of new hydration products, despite the eventual existence of unreacted cement left in old concrete [19]. The possible reactivation of the hardened cement was a relevant finding and many authors started to explore this idea regarding the production of an innovative recycled binder. As mentioned, if thermoactivation temperatures are set to be under the decarbonation stage, the CO_2 emissions may be significantly reduced, and a low-carbon binder is obtained [20]. Moreover, retrieving the waste concrete highly encourages the valorization reuse of CDW, the reduction of natural resources depletion and the relevant decrease of landfill disposal. However, recycled cement is still a very young domain of research and various aspects related to its production process and its behavior when incorporated in cement-based materials must be further explored before the implementation of this very promising eco-efficient solution in the construction industry.

The objective of this chapter is to review some of the most relevant research and main contributions achieved in the domain of thermoactivated recycled cements. First, a general overview of the recycled cement manufacture is presented. Then, Sections 3 to 5 are dedicated to the phase development in anhydrous RC and subsequent rehydration, Sections 6 to 8 discuss the main physical, microstructural and mechanical properties of cement-based materials. Section 8 also covers the use of RC as partial cement replacement, showing the higher potential of this new recycled binder compared to current mineral additions used by the concrete industry.

2. Overview of recycled cement production

The first published work concerning the specific recovery of cementitious materials regarding cement recycling was probably presented by Splittgerber and Mueller [21]. The authors suggested exploring the inversion of cement hydration based on previous studies concerning the rehydration of hardened concrete subjected to fire temperatures. In fact, as mentioned, various studies have for long been carried out in this domain [16–18]. Basically, the thermal activation turns the cement hydration into a reversible process, obtaining dehydrated compounds with similar characteristics to those of the original clinker phases [21]. As better discussed in Section 3, the thermal activation explores the chemical transformations of the cement paste that occur at different temperatures, namely involving dehydration, dehydroxylation and decarbonation stages [22–26]. The production of recycled cement essentially involves three relevant steps, namely: the separation of the cement fraction from the other constituents of waste concrete; the comminution of the waste cement to an average particle size in the range of common ordinary Portland cements (OPC); thermoactivation of waste cement into RC. The closed circular economy involved in RC manufacture is illustrated in **Figure 1**.

One main obstacle that has hindered the production of recycled cement at an industrial scale is related to the individualization and separation of waste concrete constituents. This is not only a challenge for the production of RC, but also for obtaining high quality recycled aggregates. In fact, the contamination of aggregates by adhered cement paste increases their porosity and waste absorption [27], with repercussions on the durability, shrinkage and control of fresh concrete workability [28, 29]. For this reason, current recycled concrete waste is essentially reused as low-quality recycled aggregates for low grade concretes or as backfilling in road base layers and landscape recovery. Therefore, in order to enhance the CDW valorization and encourage a closed circular economy for concrete

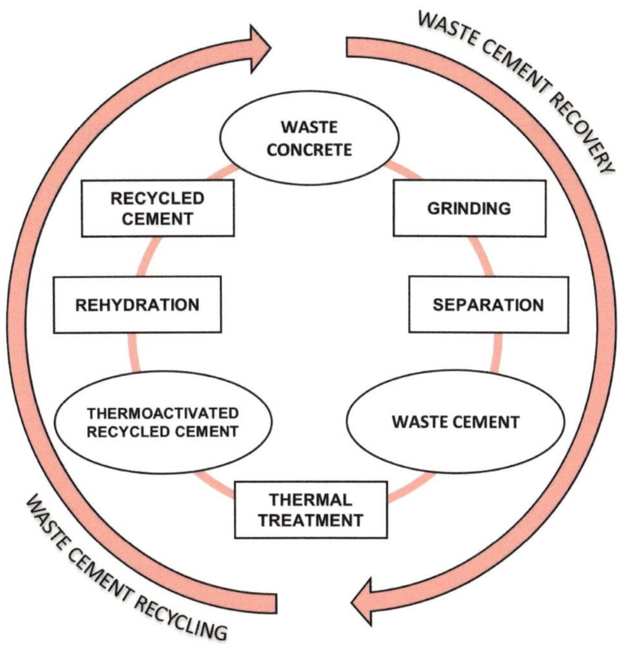

Figure 1.
Closed circular economy of thermoactivated recycled cement retrieved from hardened waste cement-based materials (adaptation from Carriço et al. [15]).

production, the development of an efficient separation process is a priority goal. However, despite various attempts, essentially focused on the recovery of cleaner recycled aggregates [30–32], no effective solution has been achieved yet. Most suggested separation methods essentially include mechanical [33] and thermal processes [31]. These methods take advantage of the different physical properties of concrete constituents, such as the crushing strength and thermal expansion, to promote separation. However, these processes are usually high-energy intensive and high contamination levels are difficult to avoid [34]. Nevertheless, some studies have thermoactivated recycled concrete fines (RCF), up to 5 mm, obtained from these processes [33, 35, 36], but the level of contamination may be high, especially when non-siliceous aggregates are considered. Other methods based on microwave heating [37, 38] or high-voltage electrical pulse discharge [39] fail to be easily implementable at an industrial scale. The lack of an effective method for waste concrete separation explains why most studies regarding the characterization of RC have involved the consideration of laboratory produced cement pastes, avoiding the challenging stage of concrete separation. Recently Bogas et al. [40] have patented a novel easily implementable and cost-effective separation process that is reported to yield waste cement with less than 12% aggregate contamination, by volume, and high quality recycled sand with as low as 3% of adhered cement paste [41].

In a second stage of RC production, waste cement is subjected to gridding, usually by means of ball milling as done in the cement industry [42, 43]. Some authors opt to previously oven dry the waste cement before gridding, since it reduces the baling phenomenon and wall mill adhesion [36, 44, 45]. In other studies, the RC grinding was performed after thermal activation [42, 46–49], however this turns the thermal process less effective and may lead to less homogeneous RC.

The increase of RC fineness enhances its rehydration reactivity, which leads to denser microstructures [50, 51]. Moreover, reducing the particle size of porous RC particles decreases their absorption properties. Therefore, it is recommended to produce RC with at least the same fineness range of OPC. However, this goal is not easily achieved in laboratory mills, especially when large amounts of RC are required. For this reason, most studies have considered particle sizes up to 150 μm. Zhang et al. [43] suggested the intergrinding of waste cement with slag (of higher hardness) in order to prevent the waste cement agglomeration and improve the fineness level.

The RC production conditions adopted by different authors are summarized in **Table 1**. The thermal treatment typically follows a thermal curve initiated by a heating rate of 5–10°C/min, followed by a residence time at maximum temperature and the respective cooling. It is expected that a lower heating rate will favor a more effective dehydration, but optimization of this parameter has never been reported [26]. However, the maximum treatment temperature and respective residence time are the main factors affecting the complete dehydration process at a given stage [59].

The first studies in this domain considered a wide range of thermoactivation temperatures, from as low as 200°C to over 900°C [46, 49, 60]. Later research has been focused on a narrower range, between 500 and 800°C [19, 33, 35, 43, 47, 61, 62], in order to comprise the phases of C-S-H dehydration and CH dehydroxylation, without relevant decarbonation. Optimal treatment temperatures have been reported to be in the range of 600–700°C, ensuring high rehydration ability and low thermal energy consumption [2]. The residence time has ranged from 1 to 8 hours in literature, although 2–3 hours is most often adopted [44, 46, 48, 49, 54, 55, 60]. The influence of the residence time and treated temperature on the mechanical strength of mortars produced with 25% of RC from the cement fraction of waste

Thermal treatment conditions				Refs.
Temperature (°C)	**Residence time (hours)**	**Heating rate (°C/min)**	**Cooling**	
200, 500, 800	1	10	cooling to room temperature	[35]
400, 650, 900	1	ND	accelerated cooling	[52]
750	1	ND	accelerated cooling	[48]
700, 750, 800	1.5	10	cooling to room temperature	[19]
650	4	10	accelerated cooling	[53]
600, 700, 800, 900	2	ND	ND	[44]
200, 400, 600, 800	2	5	ND	[46]
300, 500, 650	2	10	accelerated cooling	[54]
660–940	2.5	10	cooling to room temperature	[55]
700, 800, 900	1.5	10	600°C/min	[56]
120, 450, 750	8	7	cooling to room temperature	[42]
600	4	ND	cooling to room temperature	[42]
500	2	ND	ND	[57]
400, 500, 600, 700, 800, 900	5	20	cooling to room temperature	[58]
400, 450, 500, 600, 650, 700, 750, 800, 900	3	10	cooling to room temperature	[2]

ND, not disclosed.

Table 1.
Thermoactivation procedure of RC according to various authors.

mortar was analyzed by Kalinowska-Wichrowska et al. [63]. The maximum compressive strength was attained for RC treated at 650°C for 80 minutes.

The cooling procedure may occur inside the kiln at low cooling rates or may be accelerated by means of cooling devices, following the same philosophy adopted in OPC production [45, 48, 54, 61]. Serpell and Zunino [56] did not find significant changes when the cooling rate was accelerated.

3. Dehydrated thermoactivated recycled cement

After thermoactivation, the dehydrated RC particles are characterized by a porous structure and rough surface with high surface area, which is up to 15 times higher than that of OPC particles [21, 57]. In opposition, common OPC particles are non-porous and with smooth surface [42]. The helium particle density of RC tends to be slightly lower than that of OPC, ranging between about 2400–3200 kg/m^3 depending on the treatment temperature [2, 21, 33].

Dehydrated waste cement is usually characterized by the absence of tricalcium silicate (C_3S), as well as the significant presence of free lime (CaO) or calcite (**Figure 2**) [1]. In fact, considering the main hydration reaction of calcium silicate

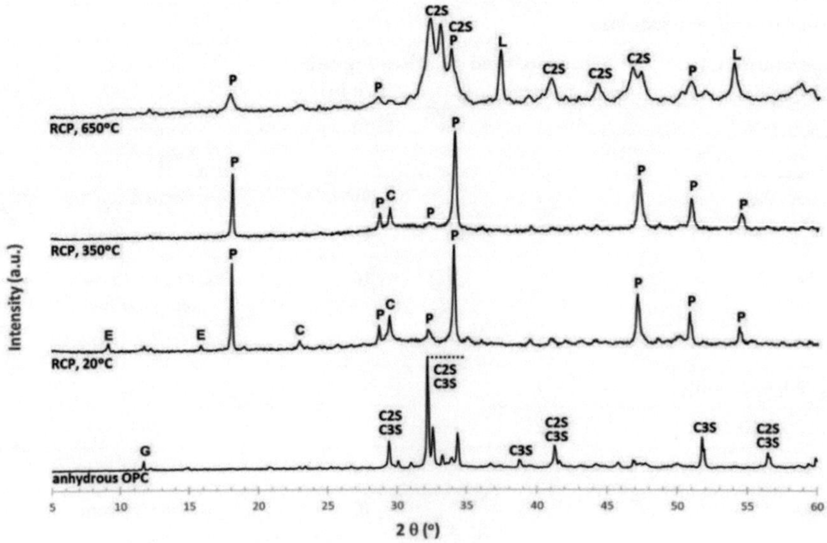

Figure 2.
XRD of anhydrous OPC, hydrated cement paste (RCP, 20°C) and RC treated at 350 and 650°C (RCP, 350°C or RCP, 650°C) [1].

compounds of OPC Eqs. (1) and (2), the obtained C-S-H may be assumed to present an average C/S ratio (CaO/SiO_2) of about 1.7 [64]. Considering the reverse dehydration of the obtained hydration products, a C/S ratio lower than 2 would be expected for the new calcium silicate as well as the presence of free clime (CaO). Based on chemical and ^{29}Si NMR analysis of RC treated at 750°C, Alonso and Fernandez [24] estimated the value of 1.78 for the C/S ratio of the new nesosilicate, concluding that the dehydrated phase coefficients pertained to a structure close to C_2S.

$$C_3S + 2.66H \rightarrow C_{1.7}SH_{1.36} + 1.3CH \quad (1)$$

$$C_2S + 1.66H \rightarrow C_{1.7}SH_{1.36} + 0.3CH \quad (2)$$

$$(H - H_2O;\ C - CaO;\ S - SiO_2)$$

The RC is also composed by other phases, such as dehydrated calcium aluminates and carbonated compounds, besides unreacted original anhydrous OPC grains and eventual residual hydration products [42, 49, 54, 58]. **Table 2** summarizes the main dehydrated phases identified by XRD in the literature.

Shui et al. [35] analyzed the dehydrated phases developed in RC treated at 200°C, 500°C and 800°C, by means of thermogravimetry (TG) and X-ray diffraction (XRD) analysis. Up to 500°C, RC was essentially composed of partially dehydrated C-S-H and CH, as well as $CaCO_3$ and C_2S, besides other amorphous dehydrated phases. Over 500°C free lime was progressively formed from the CH dehydroxylation and at 800°C the $CaCO_3$ was also decomposed into free lime. Similar conclusions regarding these major transformations with increasing temperature were also documented by other authors [1, 2, 42]. However, Wang et al. [42] reported a higher content of $CaCO_3$ after treatment at 450°C than in the source cement paste. The increase of carbonated products after thermoactivation was also documented by other authors [24, 46, 52, 65], attributing this phenomenon to the

Temperature (°C)	Precursor	Dehydrated phases	Refs.
200	RCF	C-S-H, CH, C_2S, $CaCO_3$, 4CaO.Al_2O_3Fe2O_3	[35]
500		CH, CaO, C_2S, $CaCO_3$, 4CaO.$Al_2O_3Fe_2O_3$	
800		CaO, C_2S, $CaCO_3$, 4CaO.$Al_2O_3Fe_2O_3$	
400	CP	β-C_2S, CH	[52]
650		β-C_2S, CaO	
900		β-C_2S, CaO	
600	AAC	quartz, tobermorite, CaO, β-C_2S, $CaAl_2$Si2O_8	[44]
700		quartz, tobermorite, β-C_2S, $CaAl_2Si_2O_8$, CaO, CS	
800		quartz, β-C_2S, $CaAl_2Si_2O_8$, CaO, CS, α-C_2S	
900		quartz, β-C_2S, $CaAl_2Si_2O_8$, CaO, CS, α-C_2S	
200	CP	CH, $CaCO_3$, C_2S, C_4AF	[46]
400		CH, $CaCO_3$, C_2S, C_4AF	
600		CH, $CaCO_3$, anhydrite II, C_2S, C_4AF	
800		CH, $CaCO_3$, anhydrite II, C_2S, C_4AF	
660–940	CP	α'-C_2S, β-C_2S, $CaCO_3$, C_4AF, CaO	[55]
700	CP	α'$_H$-C_2S, α -C_2S, CH, $C_{12}A_7$, C_4AF, $CaSO_4$	[56]
800		α'$_H$-C_2S, β-C_2S, CaO, $C_{12}A_7$, C_4AF	
900		β-C_2S, α'$_H$-C_2S, CaO, $Ca_5(SiO_4)_2SO_4$, C_4AF	
120	CP	CH, 1.2 nm tobermorite, jennite, $CaCO_3$	[42]
450		0.96 nm tobermorite, disordered jennite, $CaCO_3$	
750		C_2S, CaO, CS	
500	CP	quartz, $CaMg(CO_3)_2$, $CaCO_3$, C_2S, CH	[57]
400	CP	CH, tobermorite	[2]
450		CH, $CaCO_3$	
500		CH, $CaCO_3$	
600		CH, C_3A, α'$_L$-C_2S, CaO	
650		CH, C_3A, C_4AF, α'$_L$-C_2S	
700		CH, C_4AF, α'$_L$-C_2S, $CaCO_3$	
750		CH, C_4AF, α'$_L$-C_2S, CaO	
800		CH, C_4AF, α'$_L$-C_2S, β-C_2S, CaO	
900		C_4AF, α'$_L$-C_2S, β-C_2S, CaO	

RCF, recycled concrete fines; CP, cement paste; AAC, autoclave aerated concrete.

Table 2.
Dehydrated phases of RC identified through XRD according to various authors.

partial carbonation of dehydrated CH at intermediate temperatures. In addition, a higher content of carbonation products in RC is expected, because old concrete is prone to carbonate [2, 58].

Naturally, the free lime content tends to increase with the treatment temperature, especially after the decarbonation stage [25, 66]. The free lime must be taken into account during RC rehydration, since it contributes to the increase of heat release, consumption of mixing water and false setting [2, 55, 67]. Due to the high hydration susceptibility of free lime, its previous partial hydration has been reported during the stages of cooling or storage [2, 58]. It was also suggested that these newly formed CH presented lower bonding energy than the original CH in waste cement.

Wang et al. [42] reported the presence of tobermorite and jennite within the temperature range 120–450°C. The former gradually dehydrated with increasing temperature, reducing the C-S-H layer spacing from 1.2 nm to 0.96 nm up to 450°C. Above 450°C, the diffraction peaks of tobermorite and jennite disappeared, suggesting the full depolymerization of C-S-H. At 750°C poorly crystalized peaks of wollastonite (CS) and larnite (C_2S) were identified, both presenting lower reactivity than partly dehydrated C-S-H phases [42].

In another study, Lü et al. [52] documented the full dehydration of C-S-H at 400°C. This was confirmed through ^{29}Si NMR analysis, in which incipient Q^0 peaks started to replace the disilicates and chain silicates (Q^1, Q^2) of the C-S-H structures. Over 650°C, Q^0 peaks were predominant, indicating a relevant decomposition of C-S-H into what was termed as poorly crystallized β-C_2S. Similar findings were obtained by Alonso and Fernandez [24] in RC treated at 750°C, but the Q^0 peaks were attributed to a new nesosilicate of higher reactivity than β-C_2S. Finally, at 900°C, only sharp Q^0 peaks were identified by Lü et al. [52], suggesting the formation C_2S of low reactivity. The identification of β-C_2S peaks from XRD at 900°C was also reported by Serpell and Lopez [55], which become progressively sharper up to 940°C. However, at lower temperatures in the range 660–800°C, the presence of C_2S was attributed to the high temperature polymorph α'-C_2S, which is usually unstable at room temperature. The formation of this polymorph of higher reactivity than β-C_2S allowed to explain the better performance of RC treated at this temperature range than over 900°C, as also found by Shui et al. [49]. Based on XRD Rietveld analysis, Serpell and Zunino [56] later explored the development of different C_2S polymorphs in thermoactivated RC, for the range 600–800°C. According to the authors, the fraction of β-C_2S increased with the treatment temperature, substituting the polymorphs α'_H-C_2S and, with less relevance, γ-C_2S. The low temperature formation of α'_H-C2S was attributed to the direct decomposition of C-S-H into this polymorph. These new polymorphs of higher surface area and lower crystallite size presented higher reactivity [56].

Considering a wide range of RC treated between 400°C and 900°C, Real et al. [2] confirmed the partial dehydration of C-S-H to tobermorite 9 Å up to 500°C, as well as the absence of ettringite. Up to this temperature, only incipient C_2S peaks were detected by XRD, indicating the onset of C-S-H transformation into C_2S crystalline polymorphs. After 600°C, the intensity of these peaks was increased and the C-S-H depolymerization was confirmed through ^{29}Si NMR analysis. The new nesosilicate form was identified as possible α'_L-C_2S or α'_H-C_2S. Finally, above 800°C, the C_2S peaks were more intense and sharper, indicating a higher crystallinity and β-C_2S was progressively formed, suggesting the generation of a less reactive product as also reported by other authors [52, 55].

The influence of the residence time and cooling rate in the RC thermoactivation has barely been studied. Based on a surface response model, the influence of these parameters on the formation of C_2S polymorphs was analyzed by Serpell and Zunino [56]. For cooling rates ranging 187°C/h to 1925°C/h only a slight reduction of strength development was found at higher rates. The formation of dehydrated phases was significantly affected by varying the residence time between 40 and 130 minutes, increasing the fraction of α'_H-C_2S for RC treated at nearly 700°C. However, for higher temperatures, this factor assumed less relevance.

The phase composition of dehydrated RC is also affected by the composition of the precursor source material. From XRD analysis, Vyšvařil et al. [46] analyzed the relative amount of crystalline phases in different cement pastes (only OPC; 20% ground granulated blast furnace slag (GGBFS); 20% fly ash (FA)) subjected to dehydration temperatures between 200°C and 1200°C. In all samples C_2S was identified and mayenite appeared in GGBFS and FA pastes treated over 800°C. Calcium sulfoaluminate AFt and AFm phases were not identified at any temperature and treated material. However, these compounds are expected to be present in an amorphous form. This was observed by Baldusco et al. [57] from SEM images.

In sum, these studies suggest the development of a more reactive calcium silicate polymorph at intermediate temperatures, which becomes less reactive for temperatures higher than about 800°C. Overall, optimal thermoactivation temperatures are defined for the range 600–800°C. The final complex composition of RC depends on the thermal activation curve and the precursor source waste material [46, 49]. Other factors, such as the cooling rate and residence time may also affect the phase composition of RC, but further research is needed [56]. There is still some uncertainty in the characterization of the phase composition of anhydrous RC, which is affected by the different thermoactivation procedures and the limitation of the available techniques in identifying and quantifying the formed compounds (**Table 2**).

4. Rehydration of RC

The hydration mechanism of RC has yet to be fully understood. As discussed in Section 3, the dehydrated phase composition depends on the treatment temperature of RC, thus affecting the phase composition after rehydration. From XRD analysis, Xuan and Shui [60] characterized the crystalline phase composition of hydrated RC treated up to 800°C. The main phases identified in common hydrated OPC were also confirmed in RC, namely CH, $CaCO_3$ and residual C_2S and C_4AF, but associated to broader and less intense peaks. From SEM analysis, the authors found that at 400°C the RC pastes presented a looser microstructure with fine bundles of C-S-H intermixed with visible Aft phases of ettringite. However, for high temperatures up to 800°C more rehydration products were developed, and the microstructure was densified, but also with a rough and irregular morphology.

In rehydrated pastes treated up to 600°C, Vyšvařil et al. [46] also reported the presence of CH, $CaCO_3$, ettringite, C_2S and C_4AF. However, at 800°C ettringite peaks were absent and gismondine ($CA_2S_2H_4$) was identified above 600°C. Wang et al. [42] found that CH was almost absent in rehydrated RC treated at 450°C, confirmed by TG and XRD analysis. This was compensated by the increase of $CaCO_3$ content when compared to the source hydrated paste. From SEM/EDS analysis, RC presented a distinct morphology, characterized by massive nanosized clusters of C-S-H gel and $CaCO_3$, also evidencing the presence of calcium carboaluminates resulting from the reaction between calcite and aluminates. Baldusco et al. [57], for paste thermoactivated at 500°C, also reported the development of carboaluminates. In a recent work of the authors of this chapter [67], it was also confirmed the possible development of carboaluminates in regions where the calcium/carbon ratio measured by SEM/EDS was nearly 4.

According to Zhang et al. [68] the rehydration mechanism of RC implies the C-S-H formation, attributed to the repolymerization of partly dehydrated C-S-H in the presence of Ca^{2+} and water, as well as the hydration of β-C_2S. In addition, the formation of portlandite from CaO and calcium aluminates (possible C_2AH_8) from dehydrated aluminate phases was identified by the authors.

Real et al. [2] analyzed the phase composition of RC pastes after testing different temperatures, between 400°C and 900°C, through TG, XRD and ^{29}Si NMR analysis.

In general, the authors found that the rehydration was effective for any treatment temperature, presenting a similar amount of binding water as reference OPC. However, incomplete depolymerized RC, treated up to 500°C, failed to provide significant cohesive bonding between anhydrous particles. Above 600°C, the weakly formed CH presented less binding energy than the original ones in source OPC pastes. The authors found slight differences between RC and reference OPC from TG analysis, namely related to the formation of higher amounts of carbonate/sulfate Afm phases, a lower amount of CH and a higher amount of $CaCO_3$ in RC pastes. XRD analysis also confirmed the formation of similar crystalline phases in RC and OPC pastes, supporting the adequate rehydration ability of thermoactivated waste cement, especially above 600°C. The ^{29}Si NMR analysis clearly evidenced the effective depolymerization and rehydration of pastes treated above 600°C. Rehydrated RC treated between 600 and 900°C presented mean silicate chain lengths (MCL) between 3 and 6, which indicates the formation of C-S-H with a C/S ratio over 1.2 [51, 69]. Moreover, the MCL and Q^2/Q^1 ratio were of the same order in OPC and RC pastes treated at 700–800°C, suggesting the development of the same type of C-S-H in both materials. However, the estimated coefficient of hydration was higher in RC pastes, indicating the formation of a higher volume of C-S-H, at least up to 28 days. This was attributed to the higher surface area of RC and the development of interparticle products as discussed later in Section 7. Noteworthy was the slightly higher MCL and Q^2/Q^1 ratio reported for RC treated at 600°C, indicating a higher reactivity of this product with α'-C_2S in its constitution, which was able to form a greater amount of C-S-H of longer chain length. On the other hand, above 800°C an opposite trend was found in the MCL and Q^2/Q^1 ratio, confirming the slower rehydration capacity of RC associated with the formation of less reactive C_2S polymorph.

From the above studies, it is concluded that the rehydration of RC involves the generation of C-S-H, as found in OPC, and AFt or Afm phases associated with sulfoaluminate and/or carboaluminate compounds. The type and morphology of hydrated phases depend on the treatment temperature of RC. The carbonated products tend to be more abundant in RC pastes, due to residual carbonation products in source waste cement paste and the possible carbonation that occurs after thermoactivation.

5. Kinetics of RC hydration

The hydration mechanism of RC has yet to be fully understood. As discussed in Section 3, the dehydrated phase composition depends on the treatment.

The mechanism and kinetics behind the hydration process of RC are still barely understood and subject of current research. Based on isothermal calorimetry (IC) tests, Wang et al. [42] analyzed the rehydration kinetics of RC heated at 200–1010°C. A significant initial heat release rate immediately upon contact with water was observed, followed by an induction period of a few hours and an acceleration stage with a maximum peak at about 10 hours. After the acceleration stage, the heat release rate was lower in rehydrated RC than in Ref. OPC. Among all tested temperatures, RC treated at 450°C showed the highest peak of heat release, as well as the highest cumulative hydration heat. The high reactivity for 450°C was attributed to the faster rehydration of partly dehydrated tobermorite and jennite (Section 3). This seems to have been triggered by the high calcite content that may have also provided nucleation and filler effects. Similar findings were reported by Angulo et al. [54] for RC treated at 500°C, then presenting less reactivity at 650°C.

As discussed in Section 4, other studies have shown higher RC reactivity for treatment temperatures above 600°C.

Carriço et al. [58] also found that similarly to OPC, the hydration mechanism of RC treated at 700°C involves four main stages corresponding to the initial hydrolysis, induction, acceleration and deceleration. However, RC shows a remarkably higher initial heat rate, especially up to 90 minutes, in which the heat release was nearly three times higher than that of reference OPC. This phenomenon was in line with the observation of a lower setting time in RC. However, the long-term heat release of RC tended to be lower than that of OPC, suggesting the formation of less hydrated products over time.

Real et al. [2] analyzed the hydration heat evolution of RC treated between 400 and 900°C through IC analysis. The maximum heat release and acceleration peak were attained for RC treated at 600°C. The authors also observed that RC required longer periods to initiate the acceleration stage and reach the maximum heat peak than OPC, especially those cements treated above 800°C. Moreover, the RC heat release during the acceleration stage was significantly lower than that of OPC, regardless the treatment temperature. RC treated at 400°C had no significant long-term reactivity, explained by the low depolymerization level of this product. The low reactivity of RC treated at 900°C was related to the low reactivity of the C_2S polymorph obtained at this temperature. Overall, it was concluded that RC pastes tended to present longer induction periods and less intense acceleration stages.

In all these studies [2, 42, 54, 58] the IC analysis were carried out after external mixing, preventing the determination of the heat released during the first few minutes of hydration. This initial higher hydration heat release of RC was attributed by various authors to the exothermic reaction of CaO [36, 42, 54]. However, according to Baldusco et al. [57] the high initial hydration cannot be solely attributed to the hydration of CaO, because high hydration rates are observed even in RC treated at as low as 500°C, in which the amount of free lime is not as significant. The authors suggest that as stated by Shui et al. [49], this phenomenon is better explained by the high surface area and instable nature of dehydrated RC phases, which are ready to repolymerize, despite the high calcium aluminate content of RC.

Taking into account the thermoactivation at 500°C of a waste binder composed by OPC and GGBFS, Balsusco et al. [57] documented a different shape for the IC curve. In this case, the curve was characterized by a high initial peak of heat release followed by a continuous deceleration, without the identification of typical induction and acceleration stages. According to the authors, instead of a dissolution-precipitation mechanism, the rehydration followed an instantaneous re-adsorption of water. These apparently contradictory results may be explained by the partly dehydrated state of RC anhydrous phases at 500°C. Indeed, the same behavior was found by Real et al. [2] for RC treated at 400°C. As the IC analyses were prepared with internal mixing, allowing to record the heat release upon contact with water, the cumulative heat was higher in RC than in OPC. Zhang et al. [43] also recorded the initial heat release of RC treated at 600°C, reporting that this was 10 times higher than in Ref. OPC paste. In this case, the addition of gypsum was not successful in reducing the initial reactivity of RC. In another study, Zhang et al. [68] tested other set retarders in order to control the hydration evolution of RC. For RC treated at 500°C the authors found that 1–1.5% of sodium borate was able to reduce the initial peak of heat release and extend the initial setting time.

In sum, although it is generally accepted that RC is responsible for a significant initial heat release upon contact with water, the kinetic mechanisms and involved rehydration reactions are still under debate.

6. Fresh properties of recycled cement

Two main obstacles that have been hindering the application of RC are its high water demand [43, 46, 48, 53] and fast setting time [49, 58, 67]. The high water demand is essentially attributed to the high surface area [57] and porous nature [2, 43, 48, 58] of RC, and in a second plan to the free lime content [49, 53, 55, 60] and particle agglomeration [53]. Therefore, for a given workability the RC pastes usually require a high water/binder (w/b) ratio.

Shui et al. [49] found that the amount of water required for normal consistency increased from 0.48 to 0.68 with increasing treatment temperature between 300°C and 900°C. These values were about 1.8 to 2.5 times higher than those obtained for OPC. The free lime and high surface area of RC were the main reasons attributed to this phenomenon. Both factors increased with increasing temperature. Xuan and Shui [60] studied the influence of the source paste composition (w/c ratio) on the water demand of RC treated at 200–900°C. The authors found that increasing the w/c of the source paste from 0.3 to 0.5 led to a 10–20% increase of the RC water demand, regardless of the treatment temperature. This may be explained by the higher porosity and higher amount of dehydrated products, with subsequent increase of the specific surface area, in RC treated from high w/c waste paste.

Yu and Shui [53] confirmed the great propensity for RC particles to agglomerate, which is encouraged by their high surface area. The particles are agglomerated by capillary action, trapping the mixing water between them. Therefore, part of the mixing water becomes unavailable and the water demand is increased for a given workability. After sonication, the authors found a slight increase of workability due to the partial dispersion of RC particles.

However, the authors found that the water demand was not significantly affected up to 50% replacement of OPC with RC treated at 700°C. The reported increase in water demand was 1.4 and 2.5 times higher for 50% and 100% RC incorporation, respectively. According to the authors, the lubrication effect provided by the fine OPC particles aided the reduction of the friction between coarser particles, partly compensating the adverse effect of RC. In a later study, Real et al. [2] found that RC treated between 400°C and 900°C may present 2 to 3 times higher water demand than OPC pastes of equal normal consistency, increasing with the treatment temperature. A more significant increase was found above 800°C, which was attributed to the higher free lime content, after decarbonation.

In order to reduce the effect of free lime on the water demand, Serpell and Lopez [55] suggested a two-stage mixing procedure. First, water is added in order to slake the free lime and the additional water is added to compensate the water loss by evaporation, which is measured by weight difference. Nevertheless, the reported increase for RC heated between 650 and 850°C was lower than 2%, which is not significant. Similar values were reported by Carriço et al. [58]. It was concluded that the surface roughness and porosity of RC were the most influential parameters.

Most studies report lower initial and final setting time in RC than in OPC [1, 35, 43, 48, 49, 55, 58, 60, 62, 68]. The initial setting time may be as low as less than 20 minutes, which makes it less viable for building applications [49, 58]. The low setting times in RC are associated to the fast rehydration of RC, owed to their higher surface area and higher reactivity [43, 49, 57].

The low setting time is also attributed to the free lime content [24, 46]. According to Vyšvaril et al. [46] this may explain the decrease of setting time with increasing dehydration temperature. In addition, Serpell and Lopez [55] suggested that the apparent rapid setting of RC is related to a "false setting" phenomenon that may disappear after a while. A similar phenomenon is confirmed by Carriço et al. [58] for RC treated over 700°C.

The reduction of the setting time with the treatment temperature up to 800°C has been reported by various authors [43, 49, 57, 60]. Shui et al. [49] found that the initial setting time decreased from 42 to 17 minutes when the treatment temperature was increased from 300–800°C. This phenomenon was attributed to the quick repolymerization of new C-S-H, which does not involve the dissolution-precipitation process of OPC. Carriço et al. [58] found that the potential setting time of RC treated at 900°C was 2 times longer than that treated at 600°C, confirming the lower reactivity of RC for high dehydration temperatures. Similar findings were reported by Real et al. [2]. The authors also found that the potential replacement of OPC with up to 50% RC only slightly affected the setting time (less than 10%). However, Yu and Shui [53] reported a progressive reduction of the setting time with the incorporation of up to 30% RC. The same trend was documented by Yu et al. [61], for replacement percentages between 5 and 25%.

However, less expected high setting times in RC than in OPC were found by some authors [1, 2, 46]. The authors attributed this phenomenon to the possible agglomeration of RC, as well as the eventual pre-hydration of RC during cooling and storage. Setting times over 2 times higher in RC than in OPC are reported by Real et al. [2].

Taking into account source cement pastes of distinct w/c, Xuan and Shui [60] found lower setting times for high w/c cement pastes, attributed to the greater amount of dehydrated phases.

As well known, the setting time of OPC pastes is governed by the dissolution rate of C_3A and its combination with added calcium sulphate [51, 70]. However, in RC the lack of knowledge regarding the morphology of the dehydrated aluminate phases at different temperatures and their reaction mechanism does not allow to consider the same reasoning. As mentioned in Section 5, Zhang et al. [68] explored the use of different set retarders in RC treated at 600°C. The authors found that the addition of 4% gypsum only slightly increased the setting time. It seems that the dissolution of gypsum in RC can be slower than the reaction involved in the rehydration of dehydrated compounds. Contrary to OPC, in which the gypsum dissolution prevents the formation of CAH from C_3A, its addition showed to be ineffective in RC [15, 68]. Nevertheless, Sun et al. [62] documented a 6-fold increase of the setting time when 8% gypsum was added to RC thermoactivated at 550°C.

The above studies underline the challenging task of controlling the fresh behavior of RC pastes, still without an efficient solution to overcome their high water demand and non-standard setting time. Further research is needed in this domain, contributing to higher confidence in using RC.

7. Microstructure of RC pastes

Few works have been published concerning the microstructure characterization of RC cement-based materials. Most studies only involved simple qualitative scanning electron microscopy (SEM) analysis [55, 57, 68], where the phase morphology and global porosity were poorly assessed and only for a limited range of temperatures.

From SEM analysis, Shui et al. [49] documented the morphology of rehydrated phases of RC treated at 600°C, as fine bundles of C-S-H intermixed with honeycomb shape structures over the surface of dehydrated RC particles. In opposition, C-S-H in OPC presented a typical foil and fiber morphology. Overall, the RC pastes presented a looser structure than reference OPC pastes.

Zhang et al. [43], for RC treated at 600°C, also reported a highly porous hydrated paste with a weak bonding between phases. In a later study, Baldusco et al. [57] etermined the total porosity of pastes produced with RC treated at 500°C from blended OPC and GGBFS. Based on water saturation-vacuum tests it was found that RC pastes developed higher total porosity at 7 days than reference OPC pastes. This was essentially explained by the higher intrinsic porosity of dehydrated cement particles. Noteworthies are two recent studies developed by Bogas et al. [67] and Real et al. [2], where the microstructure of RC pastes was deeply characterized over time and for different dehydrated temperatures, respectively.

Bogas et al. [67] analyzed the microstructure and phase evolution of cement pastes with RC thermoactivated at 700°C, by means of TG, XRD, SEM, quantitative backscattering electron (BSE), mercury intrusion porosity (MIP) and nitrogen adsorption (NA) analysis. Tests were performed in pastes with 8, 14 and 24 hours, as well as after 3,7 and 28 days. The RC showed effective rehydration with the formation of C-S-H, AFm, AFt and other carbonation phases since early ages. Contrary to common OPC, AFm phases were developed, at least since 8 hours, associated with the formation of sulfoaluminates and carboaluminates. It was confirmed the progressive increase of CH and C-S-H over time, which shows that the hydration mechanism extends in time. The authors described the RC paste microstructure as a "dual structure", where the intraparticle porosity is surrounded by a bulk interparticle matrix, in which the available space is lower. Basically, part of the mixing water is retained by inner hydration products, reducing the amount of water and w/b ratio between RC particles. This leads to a greater proximity between RC particles, increasing the packing density and paste cohesion, and consequently the early mechanical strength. Compared to OPC pastes, RC pastes showed higher reactivity up to 3 days, higher volume of hydrated products and more refined porosity, especially at early age. However, at 28 days the compressive strength was 32% lower in RC pastes, with less development of long-term interparticle hydration products. The total porosity was only slightly lower in RC and OPC pastes of equal w/b. The higher final setting time of RC was attributed to the very early formation of interparticle hydration products that do not contribute to the cement paste cohesion.

Following this work, a comprehensive study was carried out by Real et al. [2], where the influence of the dehydration temperature, between 400°C and 900°C, on the hydration and microstructure of RC pastes was analyzed by means of TG, XRD, ^{29}Si NMR, MIP and SEM analysis. Due to the above mentioned "dual microstructure" of RC pastes, a denser microstructure was found for RC pastes tested over 600°C than in Ref. OPC pastes. However, due to the insufficient repolymerization, pastes tested up to 500°C developed very loose microstructures associated with coarse porosity and low mechanical strength. On the other hand, RC treated above 800°C was less reactive, resulting in a less dense C-S-H network. It was thus concluded that the optimal treatment temperature was within the 600–650°C range. Similar to the previous work, the compressive strength was higher in RC pastes than in OPC pastes until 3 days and 27% lower at 28 days.

8. Mechanical properties of cement-based materials with RC

Various studies have already been carried out concerning the mechanical characterization of pastes or mortars produced with incorporation of 100% RC. For different dehydration temperatures and a wide range of w/b ratios the reported compressive strength varied as much as between about 4 and 30 MPa (**Table 3**). Besides these factors, the compressive strength is also affected by other parameters, such as the characteristics of the precursor material, grinding fineness and agglomeration

Thermal treatment conditions			Cementitious material				Refs.
Precursor material	Temperature (°C)	Duration (min)	Type	w/c	$f_{cm,28d}$ (MPa)	$f_{c,Rc}$/ $f_{c,OPC}$	
CP	500	60	paste	0.40	8.0	ND	[35]
CP	800	150	paste	0.64	19.5	ND	[49]
CP	800	150	paste	0.50	20.0	ND	[60]
AAC	700	120	paste	0.45	4.2	ND	[44]
CP	800	120	paste	0.75	16.5	ND	[46]
CP	755	140	paste	0.70	31.2	ND	[56]
CP	500	120	paste	0.60	18.0	68%	[45]
CP	450	480	paste	0.55	32.3	85%	[42]
CP	650	180	paste	0.72	19.2	72%	[2]
RCP	500	60	mortar	0.26	4.7	ND	[35]
C	750	60	mortar	0.50	12.3	21%	[48]
CP	800	150	mortar	0.93	11.6	ND	[55]
CP	500	120	mortar	0.48	7.6	18%	[54]
CP	600	180	mortar	0.50	30.3	62%	[43]
CP	650	180	mortar	0.68	8.3	19%	[1]
C	650	180	mortar	0.81	0.9	2%	[1]
CP	700	300	mortar	0.58	21.6	54%	[58]
CP	650	180	concrete	0.65	33.2	83%	[2]

ND, not disclosed; RCF, recycled concrete fines; CP, cement paste; AAC, autoclave aerated concrete; C, concrete.

Table 3.
Maximum values for the 28-day compressive strength of 100% thermoactivated recycled cement pastes, mortars and concrete reported in the literature.

level of RC. When available, the relative compressive strength of RC versus that of reference OPC of equal w/b ($f_{c,Rc}/f_{c,OPC}$) is also indicated in **Table 3**. In all cases, the 28 days compressive strength of RC tended to be lower than that of OPC, confirming the development of a less dense net of interparticle hydration products, as discussed in Section 7.

When RC was directly thermoactivated from recycled concrete fines (RCF), the 28 days compressive strength was under about only 10 MPa (**Table 3**). This is related to the lack of effective separation methods allowing the increase of rehydratable compounds in the precursor material. Therefore, the development of new methods for the efficient individualization of waste concrete constituents is a priority goal in waste cement recycling. Following the new patented method of Bogas et al. [40], Real et al. [2] could retrieve cement paste from waste cement with only up to 12% aggregate contamination by volume. In this case, concrete produced with up to 30% incorporation of RC from waste concrete showed similar behavior to RC directly obtained from pure laboratory waste cement paste of equal composition.

As mentioned, most authors have opted to conduct their works using lab-made well-hydrated paste as precursor materials, in order to avoid the complex stage of concrete separation. Moreover, the full potential of RC is better accessed through non-contaminated waste precursors.

Xuan and Shui [60] showed that the mechanical strength of RC pastes is also affected by the w/c of the waste material. It was found that higher strength would be obtained using waste cement pastes with lower w/c, of 0.3 instead of 0.5. This was attributed to the higher amount of unhydrated cement particles in less hydrated low w/c pastes, which may contribute to the subsequent development of more hydrated products. Another reason may also be related to the possible production of more interparticle hydration products in these systems with less porous RC obtained from low w/c pastes.

Different trends may be found in the literature regarding the evolution of the compressive strength as a function of the treatment temperature. In some studies, the maximum mechanical strength was attained for dehydration temperatures around 500°C [42, 45, 54]. For an optimal temperature of 450°C, Wang et al. [42] explained the obtained highest strength by the quick rehydration of partially dehydrated tobermorite and disordered jennite. For higher temperatures, the presence of wollastonite and crystalline larnite reduced the subsequent reactivity of RC. However, in most cases the optimal mechanical strength has been reported to be attained within the range of 600–800°C [19, 44, 49, 56, 58, 60].

A maximum compressive strength of as high as 32 MPa was achieved by Serpell and Zunino [56] for RC treated at 750°C. As discussed in Section 3, at this temperature range the authors identified the formation of a more reactive polymorph form of dicalcium silicate (α'_H-C_2S), which progressively turned into the less reactive α-C_2S at higher temperatures. From various studies, including those exploring the use of ^{29}Si NMR analysis to characterize the structure of C-S-H in RC [2, 52], it seems reasonable to conclude that below 600°C the phase dehydration is incomplete and at higher levels the reactivity of the anhydrous Q^1 phases are highly dependent on their morphology and crystallinity. It also seems evident that over a maximum optimal treatment temperature the compressive strength is reduced [2, 42, 46, 49, 54, 56]. In fact, various studies in the literature suggest that above about 800°C the RC dehydrated phases react slowly and compressive strength is reduced [2, 42, 49, 55].

Lü et al. [52] first demonstrated that at 900°C the dehydrated SiO_2 tetrahedrons of α -C_2S were not significantly repolymerized upon water contact. Similar findings were obtained by Real et al. [2].

Bogas et al. [1] showed that the compressive strength of RC mortars might be increased when the maximum particle size of RC is reduced from 250 μm to 63 μm. Similar findings were obtained by Letelier et al. [36], comparing the performance of mortars with RC of 150 μm and 300 μm. However, taking into account RC with up to 75 μm or 150 μm, the authors did not find significant differences in the compressive strength. One reason for these differences is attributed to the agglomeration state of RC particles in the upper particle size range.

The agglomeration issue related to fine RC particles is documented by various authors [1, 2, 43]. From SEM analysis, Shui et al. [49] confirmed the poor dispersion of fine RC. Then, the same authors [53] demonstrated that the compressive strength could be almost doubled when RC was previously dispersed through sonication.

As discussed in Section 6, the increase of the treatment temperature increases the w/c ratio needed for a given workability. Naturally, this leads to a reduction of the compressive strength when compared to reference OPC of equal workability. In fact, Real et al. [41] reported a reduction of about 15% in the compressive strength when only 15% OPC was replaced with RC treated at 650°C. To compensate this, high dosages of superplasticizer have been considered in RC cement-based materials [2, 33, 61, 68]. According to Real et al. [41], the superplasticizers (SP) are also effective in reducing the mixing water in RC concrete, but the SP saturation point tends to be higher due to the porous nature and high surface area of RC. Nevertheless, the authors report that when SP is adopted, a better dispersion is attained, and a high percentage of RC may be incorporated without significantly affecting the mechanical strength.

Regarding the strength evolution, there is an almost general consensus that the hardened properties are developed faster in RC than in OPC [41, 43, 57, 67]. Shui et al. [49] found that pastes with RC treated between 400°C and 800°C showed hydration degrees at 1, 3 and 28 days of 70, 80 and 90%, respectively. Moreover, the compressive strength of RC pastes at 3 days could be higher than that of reference OPC pastes. After this point, the relative strength of RC progressively decreased to 60% of that of OPC, at 28 days. Similar findings were obtained by Bogas et al. [67] and Real et al. [2]. As mentioned in Section 7, the compressive strength at 3 days was similar to higher in RC pastes than in OPC pastes but was about 30% lower at 28 days.

On the other hand, Baldusco et al. [45] reported that the 3 days compressive strength of RC paste could be about twice that of OPC. This was explained by the hydration of residual unhydrated calcium silicate grains, as well as the fast rehydration properties of RC. After 3 days, the strength evolution was not significant, being only 6% higher at 28 days. As discussed in Section 7, the authors suggested the formation of a long-term looser microstructure in RC pastes than in OPC pastes. According to Zhang et al. [43] in RC pastes, the compressive strength at 28 days is limited by the weaker RC particles of porous nature and low hardness. However, Bogas et al. [1] suggested that the strength evolution may be severely affected by the size and agglomeration state of RC particles, explaining the delayed strength evolution beyond 7 days found in RC mortars produced by the authors.

In order to access the potential use of RC as cement addition, various authors have explored various partial substitution percentages of OPC with RC. In a first study, Bogas et al. [1] analyzed the mechanical strength of mortars produced with 20–100% RC treated at 650°C. A progressive reduction of the compressive strength was observed, but up to 20% replacement the strength was similar to that of reference OPC mortars.

In mortar with 10 to 30% replacement of OPC with RCF treated at 500°C and 800°C, Florea et al. [33] confirmed the decrease of the compressive strength with increasing RCF content. Up to 10% replacement of OPC with RCF treated at 800°C, strength was little affected, but for 30% replacement, the reduction was as high as 35%. However, Xinwei et al. [48] reported more optimistic results for pastes with 40% replacement of OPC with RC treated at 750°C, leading to only a slight reduction of 12% in the compressive strength compared to reference OPC pastes. However, for 60% replacement, the compressive strength was 36% lower. Similar replacement ratios were studied by Araújo et al. [71] for RC treated at 700°C. The authors confirmed a slight reduction of the 28 days compressive strength for 40% replacement (of about 14%) but a significant decrease for 60% replacement. In another study involving pastes produced with 5–15% RC treated at 650°C, Yu and Shui [53] found maximum compressive strengths for 5% replacement, being 30% higher than that of OPC paste. The strength was further increased by about 15%, when the RC dispersion was improved by the addition of ethanol and a sodium-based dispersant followed by sonication.

So far, only a few studies have been published regarding the mechanical strength of concrete with RC. Letelier et al. [36] analyzed the mechanical strength of concrete produced with 5–15% of RCF treated at 400°C, 500°C and 900°C and 20 to 40% recycled aggregates. No significant compressive strength reduction, below 1%, was observed when OPC was replaced with up to 15% RCF. The strength reduction was more affected by the substitution of natural aggregates with recycled aggregates than by the OPC replacement with RC.

Carriço et al. [58] studied the influence of the incorporation percentage and dehydration temperature on the physical and mechanical behavior of RC mortars.

For replacement percentages with up to 20% RC the workability and compressive strength were not significantly affected. Mortars with RC treated at 600–800°C showed the best mechanical performance. For up to 50% incorporation, the mechanical strength was only 12 to 23% lower than that of reference OPC mortars of equal w/b. Moreover, even for a high w/b of about 0.6, mortars with 100% RC were able to attain as high as 27 MPa at 28 days. It was found that RC can be comparable to the low grade OPC of class 32.5.

Qian et al. [72] analyzed the production of ultra-high performance concrete produced with lime powder, silica fume and different replacement percentages of OPC with RC treated at 650°C. The authors found that up to 25% OPC replacement the concrete workability was only minorly affected. Regarding the compressive strength, it generally decreased with the incorporation of RC. The increased air content caused by the loss of workability was the main reason attributed to this reduction, especially for incorporation percentages over 25%. Nevertheless, at 7 days the compressive strength of concrete with up to 12.5% RC was higher than that of OPC concrete.

In a more recent study, Real et al. [41] investigated the mechanical strength behavior of concrete produced with RC treated at 650°C, obtained from waste cement paste or waste concrete. Concretes were produced with total or up to 40% RC incorporation. Up to 15% RC, workability was not significantly affected. Over this level, SP had to be incorporated in RC concrete production. Overall, the mechanical strength was not significantly affected or even increased by the incorporation of up to 40% RC. Even for 100% RC, the strength reduction was only 17%. It was thus concluded that the eco-efficient RC might have great potential as a supplementary cementitious material. Moreover, the RC concrete showed higher 3 days compressive strength than OPC concrete, regardless of the RC content. As expected, the modulus of elasticity decreased with the RC content due to its lower hardness and stiffness than OPC.

9. Conclusions

In this chapter, some of the most relevant research and main challenges concerning the production of low-carbon eco-efficient thermoactivated recycled cement were addressed. It is shown that recycled cement has a great potential for the efficient reuse of the large amount of waste concrete generated worldwide and the production for the very first time of fully recycled concrete, towards a truly circular economy. So far, recycled cement comparable to the low-grade Portland cement class 32.5 can be achieved and 28-day mortar compressive strengths over 20 MPa may be easily obtained, even considering their high w/b ratios. The main challenges that must be overcome are the high-water demand and non-standard setting times of RC. However, these issues are less relevant if RC is used as supplementary material for up to 40% replacement. In addition, other issues related to the optimization of the thermal activation process, effective separation of concrete constituents, and a deeper understanding of the dehydration and hydration process, as well as the physical, mechanical and durability behavior of recycled cement based materials need to be further addressed. However, although further research is needed, a significant step forward has been already achieved with the development of a more eco-efficient recycled binder able to respond to the very demanding environmental goals of the cement industry's road map. The aim is to supply the concrete industry with a very promising low-carbon binder, addressing the more efficient use of resources, the waste disposal issue and the decrease of the carbon footprint and associated quota.

Acknowledgements

The authors would like to thank the Portuguese Foundation for Science and Technology (FCT) for funding this research through project PTDC/ECI-COM-28308/2017. The second author also wishes to thank the financial support of FCT through scholarship SFRH/BD/146033/2019.

Author details

José Alexandre Bogas*, Ana Carriço and Sofia Real
Instituto Superior Técnico, University of Lisbon, Lisbon, Portugal

*Address all correspondence to: abogas@civil.ist.utl.pt

References

[1] Bogas JA, Carriço A, Pereira MFC. Mechanical characterization of thermal activated low-carbon recycled cement mortars. J Clean Prod 2019;218:377-389. https://doi.org/10.1016/j.jclepro.2019.01.325.

[2] Real S, Carriço A, Bogas JA, Guedes M. Influence of the treatment temperature on the microstructure and hydration behavior of thermoactivated recycled cement. Materials (Basel) 2020;13. https://doi.org/10.3390/ma13183937.

[3] United Nations. Paris Agreement. Paris: 2015.

[4] European Union. Directive 2018/851 amending Directive 2008/98/EC on waste Framework. Off J Eur Union 2018:L-150/109-140.

[5] Mehta K. Reducing the Environmental Impact of Concrete. Concr Int 2011:61-66.

[6] Schneider M, Romer M, Tschudin M, Bolio H. Sustainable cement production-present and future. Cem Concr Res 2011;41:642-650. https://doi.org/10.1016/j.cemconres.2011.03.019.

[7] Flower DJM, Sanjayan JG. Green house gas emissions due to concrete manufacture. Int J Life Cycle Assess 2007;12:282-288. https://doi.org/10.1007/s11367-007-0327-3.

[8] WBCS, IEA. Cement Technology Roadmap 2009: Carbon emissions reductions up to 2050. 2009.

[9] Kajaste R, Hurme M. Cement industry greenhouse gas emissions - management options and abatement cost. J Clean Prod 2016;112:4041-4052. https://doi.org/10.1016/j.jclepro.2015.07.055.

[10] Chen W, Jin R, Xu Y, Wanatowski D, Li B, Yan L, et al. Adopting recycled aggregates as sustainable construction materials: A review of the scientific literature. Constr Build Mater 2019;218:483-496. https://doi.org/10.1016/j.conbuildmat.2019.05.130.

[11] Gartner E, Hirao H. A review of alternative approaches to the reduction of CO2emissions associated with the manufacture of the binder phase in concrete. Cem Concr Res 2015;78. https://doi.org/10.1016/j.cemconres.2015.04.012.

[12] Mineral Products Association. MPA 2050 strategy Plan 2013:1-6. http://cement.mineralproducts.org.

[13] IEA, WBCSD. Technology Roadmap: Low-Carbon Transiton in the Cement Industry. 2018. https://doi.org/10.1007/1-4020-0612-8_961.

[14] Imbabi MS, Carrigan C, McKenna S. Trends and developments in green cSustain Built Environ 2012;1:194-216. https://doi.org/10.1016/j.ijsbe.2013.05.001.

[15] Carriço A, Bogas JA, Guedes M. Thermoactivated cementitious materials - a review. Constr Build Mater 2020;250:118873. https://doi.org/https://doi.org/10.1016/j.conbuildmat.2020.118873.

[16] Poon CS, Azhar S, Anson M, Wong YL. Strength and durability recovery of fire-damaged concrete after post-fire-curing. Cem Concr Res 2001;31:1307-1318. https://doi.org/10.1016/S0008-8846(01)00582-8.

[17] Handoo SKK, Agarwal SKK, Agarwal SKK. Physicochemical, mineralogical, and morphological characteristics of concrete exposed to elevated temperatures. Cem Concr Res 2002;32:1009-1018. https://doi.org/10.1016/S0008-8846(01)00736-0.

[18] Farage MCR, Sercombe J, Gallé C. Rehydration and microstructure of cement paste after heating at temperatures up to 300 °C. Cem Concr Res 2003;33:1047-1056. https://doi.org/10.1016/S0008-8846(03)00005-X.

[19] Serpell R, Lopez M. Reactivated cementitious materials from hydrated cement paste wastes. Cem Concr Compos 2013;39:104-114. https://doi.org/10.1016/j.cemconcomp.2013.03.020.

[20] He Z, Zhu X, Wang J, Mu M, Wang Y. Comparison of CO2 emissions from OPC and recycled cement production. Constr Build Mater 2019;211:965-973. https://doi.org/10.1016/j.conbuildmat.2019.03.289.

[21] Splittgerber F, Mueller A. Inversion of the Cement Hydration As a New Method for Identification and/or Recycling ? 11th Int. Congr. Chem. Cem., 2003, p. 1282-91. https://doi.org/10.13140/2.1.2201.3766.

[22] Monteagudo SM, Moragues A, Gálvez JC, Casati MJ, Reyes E. The degree of hydration assessment of blended cement pastes by differential thermal and thermogravimetric analysis. Morphological evolution of the solid phases. Thermochim Acta 2014;592:37-51. https://doi.org/10.1016/j.tca.2014.08.008.

[23] Alarcon-Ruiz L, Platret G, Massieu E, Ehrlacher A. The use of thermal analysis in assessing the effect of temperature on a cement paste. Cem Concr Res 2005;35:609-613. https://doi.org/10.1016/j.cemconres.2004.06.015.

[24] Alonso C, Fernandez L. Dehydration and rehydration processes of cement paste exposed to high temperature environments. J Mater Sci 2004;39:3015-3024. https://doi.org/10.1023/B:JMSC.0000025827.65956.18.

[25] Castellote M, Alonso C, Andrade C, Turrillas X, Campo J. Composition and microstructural changes of cement pastes upon heating, as studied by neutron diffraction. Cem Concr Res 2004;34:1633-1644. https://doi.org/10.1016/S0008-8846(03)00229-1.

[26] Zhang Q, Ye G. Dehydration kinetics of Portland cement paste at high temperature. J Therm Anal Calorim 2012;110:153-158. https://doi.org/10.1007/s10973-012-2303-9.

[27] de Juan MS, Gutiérrez PA. Study on the influence of attached mortar content on the properties of recycled concrete aggregate. Constr Build Mater 2009;23:872-877. https://doi.org/10.1016/j.conbuildmat.2008.04.012.

[28] Behera M, Bhattacharyya SK, Minocha AK, Deoliya R, Maiti S. Recycled aggregate from C & D waste & its use in concrete – A breakthrough towards sustainability in construction sector : A review. Constr Build Mater 2014;68:501-516. https://doi.org/10.1016/j.conbuildmat.2014.07.003.

[29] Shi C, Li Y, Zhang J, Li W, Chong L, Xie Z. Performance enhancement of recycled concrete aggregate - A review. J Clean Prod 2016;112:466-472. https://doi.org/10.1016/j.jclepro.2015.08.057.

[30] Al-Bayati HKA, Das PK, Tighe SL, Baaj H. Evaluation of various treatment methods for enhancing the physical and morphological properties of coarse recycled concrete aggregate. Constr Build Mater 2016;112:284-298. https://doi.org/10.1016/j.conbuildmat.2016.02.176.

[31] Ahn JW, Kim HS, Han GC. Recovery of aggregates from waste concrete by heating and grinding. Geosystem Eng 2001;4:117-122. https://doi.org/10.1080/12269328.2001.10541178.

[32] Sui Y, Mueller A. Development of thermo-mechanical treatment for recycling of used concrete. Mater Struct Constr 2012;45:1487-1495. https://doi.org/10.1617/s11527-012-9852-z.

[33] Florea MVA, Ning Z, Brouwers HJH. Activation of liberated concrete fines and their application in mortars. Constr Build Mater 2014;50:1-12. https://doi.org/10.1016/j.conbuildmat.2013.09.012.

[34] Akbarnezhad A. Separation processes to improve the quality of recycled concrete aggregates (RCA). Woodhead Publishing Limited; 2013. https://doi.org/10.1533/9780857096906.2.246.

[35] Shui Z, Xuan D, Wan H, Cao B. Rehydration reactivity of recycled mortar from concrete waste experienced to thermal treatment. Constr Build Mater 2008;22:1723-1729. https://doi.org/10.1016/j.conbuildmat.2007.05.012.

[36] Letelier V, Tarela E, Muñoz P, Moriconi G. Combined effects of recycled hydrated cement and recycled aggregates on the mechanical properties of concrete. Constr Build Mater 2017;132:365-375. https://doi.org/10.1016/j.conbuildmat.2016.12.010.

[37] Akbarnezhad A, Ong KCG, Zhang MH, Tam CT, Foo TWJ. Microwave-assisted beneficiation of recycled concrete aggregates. Constr Build Mater 2011;25:3469-3479. https://doi.org/10.1016/j.conbuildmat.2011.03.038.

[38] Bru K, Touzé S, Bourgeois F, Lippiatt N, Ménard Y. Assessment of a microwave-assisted recycling process for the recovery of high-quality aggregates from concrete waste. Int J Miner Process 2014;126:90-98. https://doi.org/10.1016/j.minpro.2013.11.009.

[39] Shigeishi M. Separation and collection of coarse aggregate from waste concrete by electric pulsed power. AIP Conf Proc 2017;1887. https://doi.org/10.1063/1.5003560.

[40] Bogas JA, Pereira M, Guedes M, Carriço A, Hu S, Sousa R. Separation process for obtaining recycled cement from waste concrete. 116130, 2020.

[41] Real S, Bogas JA, Carriço A, Hu S. Mechanical Characterisation and Shrinkage of Thermoactivated Recycled Cement Concrete. Appl Sci 2021;11:2454. https://doi.org/10.3390/app11062454.

[42] Wang J, Mu M, Liu Y. Recycled cement. Constr Build Mater 2018;190:1124-1132. https://doi.org/10.1016/j.conbuildmat.2018.09.181.

[43] Zhang L, Ji Y, Huang G, Li J, Hu Y. Modification and enhancement of mechanical properties of dehydrated cement paste using ground granulated blast-furnace slag. Constr Build Mater 2018;164:525-534. https://doi.org/10.1016/j.conbuildmat.2017.12.232.

[44] Shui Z, Lu J, Tian S, Shen P, Ding S. Preparation of new cementitious system using fly ash and dehydrated autoclaved aerated concrete. J Wuhan Univ Technol Mater Sci Ed 2014;29:726-732. https://doi.org/10.1007/s11595-014-0987-3.

[45] Baldusco R, Nobre TRS, Angulo SC, Quarcioni VA. Reatividade e resistência mecânica de pastas reidratadas de cimento de alto forno. Encontro Nac. sobre aproveitamento resíduos na construção, Fortaleza: 2017.

[46] Vyšvařil M, Bayer P, Chromá M, Rovnaníková P. Physico-mechanical and microstructural properties of rehydrated blended cement pastes. Constr Build Mater 2014;54:413-420. https://doi.org/10.1016/j.conbuildmat.2013.12.021.

[47] Lü L, He Y, Hu S. Binding materials of dehydrated phases of waste hardened cement paste and pozzolanic admixture. J Wuhan Univ Technol Mater Sci Ed 2009;24:140-144. https://doi.org/10.1007/s11595-009-1140-6.

[48] Xinwei M, Zhaoxiang H, Xueying L. Reactivity of Dehydrated Cement Paste from Waste Concrete Subjected to Heat Treatment. Sencond Int. Conf. Sustain. Constr. Mater. Technol., 2010, p. 175-80.

[49] Shui Z, Xuan D, Chen W, Yu R, Zhang R. Cementitious characteristics of hydrated cement paste subjected to various dehydration temperatures. Constr Build Mater 2009;23:531-537. https://doi.org/10.1016/j.conbuildmat.2007.10.016.

[50] Bullard JW, Jennings HM, Livingston RA, Nonat A, Scherer GW, Schweitzer JS, et al. Mechanisms of cement hydration. Cem Concr Res 2011;41:1208-1223. https://doi.org/10.1016/j.cemconres.2010.09.011.

[51] Kurdowski W. Cement and Concrete Chemistry. Springer; 2014. https://doi.org/10.1007/978-94-007-7945-7.

[52] Lü L, He Y, Hu S. Structural characteristics of dehydrated phase of hardened cement paste and its rehydrating ability. J Chinese Ceram Soc 2008;36.

[53] Yu R, Shui Z. Influence of agglomeration of a recycled cement additive on the hydration and microstructure development of cement based materials. Constr Build Mater 2013;49:841-851. https://doi.org/10.1016/j.conbuildmat.2013.09.004.

[54] Angulo SC, Guilge MS, Quarcioni VA, Baldusco R, Cincotto MA. Rehydration of Cement Fines: a Tg /Calorimetry Study. III Prog Recycl Built Environ 2015:222-229.

[55] Serpell R, Lopez M. Properties of mortars produced with reactivated cementitious materials. Cem Concr Compos 2015;64:16-26. https://doi.org/10.1016/j.cemconcomp.2015.08.003.

[56] Serpell R, Zunino F. Recycling of hydrated cement pastes by synthesis of α′H-C2S. Cem Concr Res 2017;100:398-412. https://doi.org/10.1016/j.cemconres.2017.08.001.

[57] Baldusco R, Nobre TRS, Angulo SC, Quarcioni VA, Cincotto MA. Dehydration and Rehydration of Blast Furnace Slag Cement. J Mater Civ Eng 2019;31:1-13. https://doi.org/10.1061/(ASCE)MT.1943-5533.0002725.

[58] Carriço A, Real S, Bogas JA, Pereira MFC. Mortars with thermo activated recycled cement: fresh and mechanical characterisation. Constr Build Mater 2020;256. https://doi.org/https://doi.org/10.1016/j.conbuildmat.2020.119502.

[59] Collier NC. Transition and decomposition temperatures of cement phases - a collection of thermal analysis data. Ceram - Silikaty 2016;60:338-343. https://doi.org/10.13168/cs.2016.0050.

[60] Xuan DX, Shui ZH. Rehydration activity of hydrated cement paste exposed to high temperature. Fire Mater 2010:481-490. https://doi.org/10.1002/fam.1067.

[61] Yu R, Shui Z, Dong J. Using dehydrated cement paste as new type of cement additive. ACI Mater J 2013;110:395-401.

[62] Sun T, Shui ZH, Huo T. Rehydration Performance of Binary Binders Made with Dehydrated Cement Paste and Phosphogypsum. Key Eng Mater 2011;474-476:1238-1242. https://doi.org/10.4028/www.scientific.net/KEM.474-476.1238.

[63] Kalinowska-wichrowska K, Kosior-kazberuk M. The Properties of Composites with Recycled Cement Mortar Used as a Supplementary Cementitious Material. Materials (Basel) 2020;13. https://doi.org/doi:10.3390/ma13010064.

[64] Richardson IG. Nature of the hydration products in hardened cement pastes. Cem Concr Compos 2000;22:97-113. https://doi.org/10.1016/S0958-9465(99)00036-0.

[65] Song H, Jeong Y, Bae S, Jun Y, Yoon S, Eun Oh J. A study of thermal decomposition of phases in cementitious systems using HT-XRD and TG. Constr Build Mater 2018;169:648-661. https://doi.org/10.1016/j.conbuildmat.2018.03.001.

[66] Zhang Q, Ye G. Quantitative analysis of phase transition of heated Portland cement paste. J Therm Anal Calorim 2013;112:629-636. https://doi.org/10.1007/s10973-012-2600-3.

[67] Bogas JA, Carriço A, Tenza-Abril AJ. Microstructure of thermoactivated recycled cement pastes. Cem Concr Res 2020;138:106226. https://doi.org/10.1016/j.cemconres.2020.106226.

[68] Zhang L, Ji Y, Huang G, Gao F, Dong Z. Effect of retarders on the early hydration and mechanical properties of reactivated cementitious material. Constr Build Mater 2019;212:192-201. https://doi.org/10.1680/jadcr.18.00051.

[69] Hou D. Molecular Simulation on Cement-Based Materials. 2020. https://doi.org/10.1007/978-981-13-8711-1.

[70] Taylor HFW. Cement chemistry. Thomas Telford; 1997.

[71] Araújo Jr. AG de, Vale AE, Azevedo AG de S, Strecker K. Estudo do reaproveitamento do cimento Portland de alta resistência inicial. Cerâmica 2017;63:58-64. https://doi.org/10.1590/0366-69132017633652052.

[72] Qian D, Yu R, Shui Z, Sun Y, Jiang C, Zhou F, et al. A novel development of green ultra-high performance concrete (UHPC) based on appropriate application of recycled cementitious material. J Clean Prod 2020;261:121231. https://doi.org/10.1016/j.jclepro.2020.121231.

Chapter 4

Different Approaches to Develop More Sustainable Concrete Alternatives

Mauricio Pradena and Andrés César

Abstract

As important as it is, sustainability related with the concrete material is more than reducing the amount of cement in concrete mixes. In effect, there can be other types of contributions to a sustainable development using this fundamental material. The purpose of this book chapter is to analyse some of these approaches, in particular, concrete durability, reducing the amount of required concrete (and then cement) through innovative structural design, and reducing the amount of aggregates used in the concrete material. More specifically, examples and results obtained in Chile with biological self-healing concrete, thinner concrete pavements and concrete with industrial and domestic waste as partial aggregate replacement are included in the chapter. Due to its importance, the geo-dependency of the concrete material is addressed as well.

Keywords: sustainability, concrete with waste, concrete durability, thinner concrete pavements, self-healing concrete, life-cycle

1. Introduction

Concrete is the most widely used building material due to its strength, adaptability, low maintenance requirements during the lifetime of structures, and the economic and extended accessibility of its components [1], which makes it very difficult to replace in many infrastructure applications [2]. In fact, globally, concrete production is estimated at approximately 25 billion tonnes per year [3].

However, a relevant challenge is the amount of harmful emissions produced by concrete [4]. Actually, cement manufacturing and concrete production account for 6–9% of global man-made CO2 [5, 6]. Furthermore, by 2050, this value is projected to increase by 4% due to an increase of 12–23% in cement consumption [7]. Cement production generates, as well, high NOx and SOx emissions, which contribute to the development of acid rain, deterioration of public health and global climate change [8, 9].

Additionally, the high demand for natural aggregates used in the concrete mix generates a loss of vegetation and fauna, a decrease in air quality due to emissions of particulate matter, loss of fertile soil, risks of contamination of groundwater and deterioration of the life quality of people living near the extraction sites [10]. Furthermore, the extraction of aggregates can cause morphological alterations in the shape of the channel, bottom and banks, which has repercussions upstream through rebound erosion, generating a considerable increase in the velocity and shear stress of the flow and favouring the erosion processes in the riverbed [11].

The negative impacts related with the concrete material can be minimised by adopting a sustainable approach [12], where sustainability is understood as meeting the needs of the present without compromising the ability of future generations to meet their own needs [13]. In this context, sustainable engineering implies not only the use of sustainable materials, but a sustainable engineering system, where different engineering stages and processes can include factors such as reduction of environmental impacts, economic accessibility and access to the engineering solution regardless of the geographical location [14]. Therefore, from this holistic approach, the result is an alternative with ae technical, environmental, economic and social balance [15].

As important as it is, sustainability related with the concrete material is more than reducing the amount of cement in concrete mixes. Actually, sustainable characteristics have been reported by optimising the design of structures [16–18], replacing natural aggregate with different types of waste such as recycled crushed concrete, marble waste [19] or waste foundry sand [20] and incorporating non-traditional materials such as fibres [21, 22], biological material [23], hazardous waste material [24] or nanomaterials into concrete [25].

The objective of this book chapter is to analyse different types of contributions to a sustainable development using the concrete material. Although some examples are mentioned for the case of reducing the amount of cement directly in the concrete mixes, the analysis is focussed on the other approaches presented in Section 3 and analysed in Section 4 using different examples, most of them proposed from the concrete laboratory of the University of Concepción, Chile. Related with this, Section 2 presents the fundamental concept of geo-dependency, which is directly related with the development of practical and useful sustainable concrete alternatives.

2. Geo-dependency of the concrete material

Concrete is composed of local raw materials. This characteristic is one of the explanatory factors of the massive use of concrete in the construction industry. However, this geo-dependency means that the specific types of cement used, the aggregates and the water influence the final behaviour of the concrete.

The aggregates represent between 70% and 80% of the volume of concrete and almost 90% of the total concrete weight. Therefore, the shape, surface roughness and mineralogical type of the aggregate plays a relevant role in defining the thermal and mechanical properties of the concrete [26].

Similarly, the composition of the cement and the fineness with which it is ground in its manufacture have significant effects on the concrete behaviour. For instance, in the heat of hydration, which can have effects on the durability and service life of concrete structures [27, 28].

Furthermore, the performance of concrete in service is also influenced by conditions of the site environment such as levels of chlorides, sulphates, humidity, pH, CO_2 and thermal amplitude.

Hence, the concrete is a geo-dependent material because its functional performance is determined by the raw materials used in its manufacture and the local construction conditions. In this way, concrete mixes produced in two different regions of the planet are different, even though they can have the same strength.

Geo-dependence is also crucial when dealing with waste materials incorporated in concrete, especially because these alternative solutions must be practical and feasible to implement in order to effectively reuse the waste [29]. Therefore, they must be feasible not only from a technical point of view, but from an economic perspective, as well, including all associated costs.

In effect, if the final purpose is to efficiently reuse waste by incorporating it into concrete, geo-dependency leads to an evaluation of the performance of these solutions according to the local context. In this way, the results obtained contribute to the optimal use of waste and then to sustainable development, which can be adopted in other regions as well. For instance, regions with similarities, that can be in the type of cement composition, cement substitute materials and/or characteristics of the aggregates.

3. Sustainable approaches related with the concrete material

3.1 Approaches based on the phases of the concrete material life-cycle

The methodology to assess the environmental impact of a material, includes the calculation of the related emissions generated and energy required in the different phases of the material life-cycle, i.e. from the extraction of raw materials to the end of the first life. Therefore, the use of the life-cycle phases of the concrete material results especially useful to analyse different contributions to more sustainable alternatives of this massive use material.

Evaluating the energy consumption and generated emissions of material alternatives allows to quantify sustainable benefits at different stages of the life-cycle. For instance, in the material production stage, alternatives can be generated to reduce the amount of cement and natural aggregates used in the concrete mix. And in the product manufacture stage, benefits can be achieved by implementing optimised design and construction procedures that reduce the amount of concrete or alternatives with less demanding maintenance interventions, or with better thermal insulation and then less demanding of heating energy. Furthermore, more durable concrete can be produced in such a way that at the end of the first life the material can be reused.

In the present section some of those different approaches are introduced (**Figure 1**).

3.2 Reducing the amount of cement in the concrete mix

An alternative to minimise the harmful emissions generated by concrete is to directly reduce the consumption of cement in the mix. For that purpose, supplementary cementitious materials have been developed to replace part of the required cement [31].

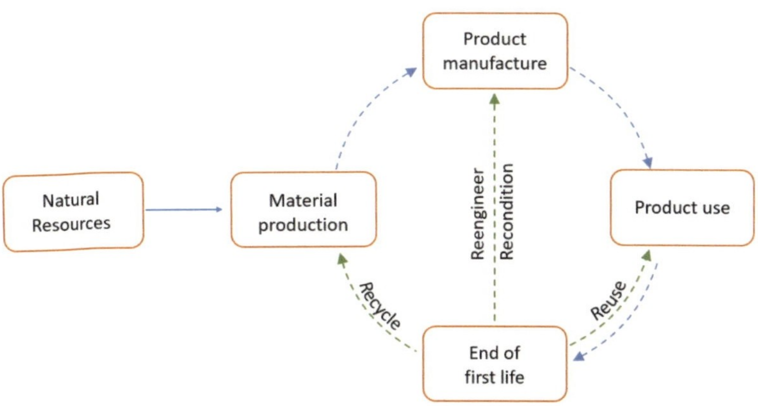

Figure 1.
Material life-cycle [30].

These alternative materials include the reuse of waste such as fly ash, which can replace 15–30% of cement [32–36] and silica fume which replace cement by 5–25% [37, 38]. There are other residues as well that have been used, such as ground granulated blast furnace slag [39], metakaolin [40, 41], sewage sludge ash [42, 43], rice husk ash [44], which can replace up to 20% of cement.

Alkaline activated cements can be used to replace more than 50% of conventional cement [45–48]. This material is a product with cementitious properties generated from the reaction of a powdery material with an aluminosilicic nature and an alkaline agent [49]. Alkaline cements are mainly characterised by low hydration heats, high mechanical performance, good durability against different chemical attacks, often using industrial waste as the only raw material and not requiring high energy consumption compared to the Portland cement manufacturing process [50].

Even a "cementless concrete" can be produced by implementing a sub-group of alkaline-activated materials called geopolymers [50]. In this regard, the main author of this chapter has the experience of design and making geopolymer material with Chilean aggregates in the Microlab TU Delft. The results were optimal, obtaining without problems the design strength.

However, the implementation of the technology of actively alkaline cements and geopolymers requires great care in their production. In addition, the polymerisation reaction is very sensitive to temperature and requires the curing of the geopolymer concrete to be at an elevated temperature under a strictly controlled temperature regime [51, 52]. These aspects coupled with the lack of standardisation of the material make the practical implementation of this technology limited to particular engineering solutions such as the construction of prefabricated structures like sewer pipes and marine members.

3.3 Implementation of optimised designs

Another approach to reduce the environmental impact of concrete is to implement sustainable design features. For instance, optimised structures requiring fewer concrete material to satisfy the in-service demands or optimised structures requiring less conditioning energy during their use phase. Optimisation has been applied in the design of hydraulic structures as dams [16–18]. Actually, Deepika and Suribabu used differential evolution algorithm to find the best optimal shape of a gravity dam reducing 20% the demand of concrete [53].

Similarly, Hashemian proposed a cambered curve beam which requires 20% less concrete than the conventional prismatic beam [54].

Other possibility of optimization is to take better advantage of intrinsic properties of the concrete material in the design of structures. An of example of is the concrete thermal inertia, which is the characteristic of taking a long time to warm up, but, once warm, taking a long time to cool down, and vice versa. This property incorporated in building design can result in structures that maintain a comfortable internal temperature by means of a better interaction with the environment [55, 56]. As a result, a reduction of the conditioning energy over the lifetime of the building is achieved.

3.4 Improving the concrete durability

From a sustainable and technical perspective, the greater the durability of the concrete material, the greater the benefits. Certainly, it is not the same to have a concrete material, for the same application, enduring 15 or 60 years. If the structure has good durability, then it will require less significant interventions and the service life can be extended. In this regard, the ICRI Committee 160 mentions that the most effective sustainability strategy for concrete and masonry structures is to avoid the need of repairing [57].

A particular case is reinforced concrete, where it is very important to avoid or reduce the entrance of atmospheric agents that can corrode the reinforcement [58–60]. This can prevent advanced deterioration and damage, which repair works, require substantial resources. For example, in the UK the annual cost of repairing reinforced concrete structures near coastal areas is £755,000,000 [61]. Similarly, between 1991 and 2001, the cost of corrosion repair of reinforced concrete structures in the United States was $276 million, representing 3.1% of the gross domestic product [62].

In order to prevent corrosion deterioration, it is essential to reduce the formation and propagation of micro-cracking generated by the concrete shrinkage phenomena at early age [60, 63, 64]. As alternatives to limit the micro-cracking and thus increase the concrete durability, Jonkers et al. developed a promising self-healing concrete technology based on the application of mineral produced by bacteria included in the mix [23]. This method allows to repair from inside the material without requiring external agents to activate the process.

Another widely evaluated alternative is the incorporation of fibres in the concrete, which can reduce the number, size and propagation of microcracks [21, 22]. In this case, compared to man-made fibres, natural fibres are less expensive, locally available (in some cases as industrial waste), renewable, lightweight, biodegradable and less energy intensive to produce [65–67].

3.5 Reducing the amount of natural aggregates in the concrete mix

The concrete industry generates a high demand for natural aggregates, due to the massive use of concrete, and the fact that approximately 70% of the volume of this material is composed by aggregates [68]. Uribe mentions that aggregate extraction processes generate loss of vegetation cover and soil [10]. Consequently, the landscape and the habitat of the existing fauna in the area are altered. Moreover, soil fertility and air quality are reduced due to emissions of particulate matter. Therefore, the life quality of the people living near the extraction sites is affected as well.

Another direct impact relates to construction costs. Indeed, as aggregates are a finite resource, they can become scarce, which can increase the cost of acquiring and transporting materials from more distant locations.

Globally, up to 17.5 Gt of aggregates have been consumed annually for concrete manufacturing [69]. In the United States, two billion tonnes of aggregates are produced annually [70]. Similarly, it is estimated that in Chile, aggregate extraction amounts to 7 million 500 thousand cubic metres produced annually, of which five million correspond to gravel and sand [71].

Therefore, in order to preserve natural resources and, at the same time, contributing to solve waste disposal problems, it has been evaluated to replace natural aggregates by different types of waste such as recycled crushed concrete, crushed bricks, recycled glass, rubber, ceramics, marble waste, textile effluent sludge [19], waste foundry sand [20], steel slag [72], copper slag [73], blast furnace slag, ferrochrome slag, class F fly ash, palm oil clinker [68] and various types of plastic waste [74].

4. Examples of sustainable concrete alternatives

4.1 Case studies

Based on life cycle approaches, the concrete laboratory of the Universidad de Concepcion in Chile has made efforts to contribute to the development of technologies that decrease the harmful impact generated by the concrete industry. This section describes these technologies and presents the relevant results of each study.

4.2 Reduction of the amount of cement by means of optimised structural pavement design

Since all the desired objectives and requirements of an engineering structure are specified in the design phase [75], in the case of concrete pavements, the structural design plays an important role in improving the sustainability of the road infrastructure [76]. In this regard, the design method of short slab Jointed Plain Concrete Pavements (JPCP) is a patented alternative that proposes to shorten the slabs size in such a way that there is only one set of wheel loads per slab [77, 78]. Therefore, the traditional configuration where the slab can support the full vehicle load does not occur. The benefits of this design include, as well, a reduced slab curvature. In this way, the pavement thickness can be reduced up to 10 cm [78], which means than less concrete is required, and then the demand of cement, aggregates and water decreases.

Additionally, it is postulated that the primary method of transferring loads between slabs is the aggregate interlock mechanism, and then, dowel bars are not part of the standard design.

These features result in savings of approximately 20% of the construction cost, which makes concrete pavements with optimised geometry a competitive solution compared to asphalt pavements when only construction costs are considered in the analysis, i.e. without life-cycle costs [78]. This is especially relevant in developing countries, where construction costs can be the main limit to the application of more durable pavements with less maintenance intervention requirements than asphalt pavements [79, 80]. In fact, promising experiences with concrete pavements with optimised geometry have been carried out in developing countries such as Chile, Guatemala, Nicaragua and Peru [78, 81–83]. Moreover, evaluations of short slab test sections in the United States indicate that these pavements have the ability to maintain in-service performance similar to traditional concrete pavements up to 51.3 million equivalent axle loads (ESAL) [84].

In order to estimate the benefit of this engineering innovation, the emissions generated in 18 short slab projects and their traditional JPCP equivalents were compared [85]. The assessment was based on environmental management principles and measures the impact on the environment and human health [86, 87]. Furthermore, it is assumed that the design hypotheses are fulfilled and then the in-service performances of both pavement alternatives are equivalent. For this reason, the evaluation is focused from the material procurement stage to the construction phase, which presents differences between the two design philosophies.

The results indicate that, for all cases evaluated, the concrete pavements with optimised geometry generate lower harmful impacts (**Figure 2**). Regarding environmental impacts, all indicators presented reductions above 10%, with global warming, acidification and ozone depletion showing the highest reductions with 33%, 27% and 24%, respectively. In terms of human health impacts, short slab pavements reduce, as well, exposure to carcinogenic, non-carcinogenic and respiratory pollutants by 26%, 23% and 27% respectively.

The reduction in these indicators reflects that concrete pavement with optimised geometry contribute to a more sustainable construction. The visible advantages of short slab pavements are related to their ability to maintain a performance similar to a traditional JPCP while using thinner slabs. This reduction means less cement consumption, which is the main generator of emissions.

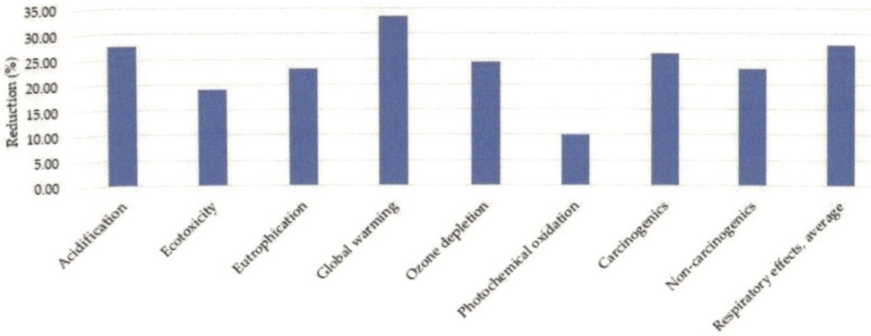

Figure 2.
Percentage of emission reduction with short slabs [85].

4.3 Increasing durability to reduce the concrete demand

4.3.1 Natural fibres to control concrete microcracking

In order to prevent internal corrosion of reinforced concrete, it is essential to reduce the formation and propagation of microcracking generated at early age by the concrete shrinkage phenomena [60, 63, 64]. Traditionally, little attention has been focused on microcracks due to their low immediate structural impact. However, from a sustainable development perspective, the control of microcracks is fundamental to limit the entrance of atmospheric agents that can corrode the rebars. Hence, the reduction of microcracks results in more durable concretes, which is a direct contribution to sustainability.

Microcracking concrete control is especially attractive in regions with aggressive environments for reinforcement corrosion, such as coastal areas. In this sense, due to its geography and the length of its coastline (6,435 km), Chile is a country that is particularly suitable for the application of these concretes. In fact, of the ten largest cities of the country, 6 are located in coastal areas.

The traditional process to limit the number and size of microcracks in concrete is the incorporation of industrial fibres [88–90]. However, abundant virgin raw materials are required to manufacture steel, glass, or plastic fibres. In this context, natural fibres have lower embedded energy, are cheaper, renewable, biodegradable and locally abundant [65–67]. In fact, they may be available as waste, which makes their use in concrete even more attractive.

Research developed by Soto et al. [91] and Okeola et al. [92] show that sisal fibres limit the propagation of microcracks in concrete. However, the incorporation of natural fibres can reduce the compressive strength. Indeed, the reduction is greater if the amount of fibres increases [93]. This has been observed in coconut fibre with reductions in compressive strength between 11% and 13% [94], jute fibre with reductions between 6% and 35% [95] and sisal fibre with reductions in compressive strength between 4.22% and 25.30% [92].

Considering that the mechanical properties, and in particular the compressive strength, is fundamental for the massive use of concrete, it is fundamental the evaluation of those properties if a particular type of natural fibre wants to be considered to control concrete microcracking. In particular, at the University of Concepción, a study has been developed in order to evaluate mechanical properties of concrete with *Eucalyptus Globulus* bark fibre, which is a waste product of the forestry industry widely available in Chile [29, 96].

A total of four fibre inclusion percentages were evaluated with respect to the weight of cement: 0.5%, 1.0%, 2.0%, and 5.0%. In order to study the potential influence of fibre absorption on the performance of the samples, the fibres were included in dry and saturated state.

The study considered, as well, the evaluation of the potential effects on the durability of the fibre when impregnated with a paraffin emulsion. This is because modification of the fibre surface with chemical or physical agents is a strategy to mitigate the potential degradation caused by the alkaline environment of the cementitious matrix (**Figure 3**).

The results indicated that there is no significant difference between incorporating the fibre in a dry or saturated state. Therefore, it is not essential to dry the fibre before incorporating it into the mixture. With respect to the fibres treated with paraffin emulsion there are no strength advantages related with their application.

In regard to mechanical behaviour, samples with 05% *Eucalyptus Globulus* bark fibre have a remarkably performance, although it is lower compared to samples without fibre. For instance, by incorporating 0.5% dry fibre (CD-0.5), the compressive strength after 28 days only differs by 0.98 MPa from the control sample without fibre (CC), which is very close to the standard error of 0.82 MPa. This trend does not agree with that reported by other researchers, which indicate significant reductions in the strength of concrete with natural fibres [91, 94, 95]. A similar behaviour was observed in the flexural strength, where the saturated samples (CS-0.5) showed a reduction of 0.20 MPa, being 0.22 MPa the standard error, i.e. the difference cannot be considered significant. Furthermore, the samples with dry fibres (CD-0.5) show a reduction of 0.7 MPa (**Figure 4**).

Figure 3.
Samples with E. globulus *bark fibres: (a) top view; (b) side view [29].*

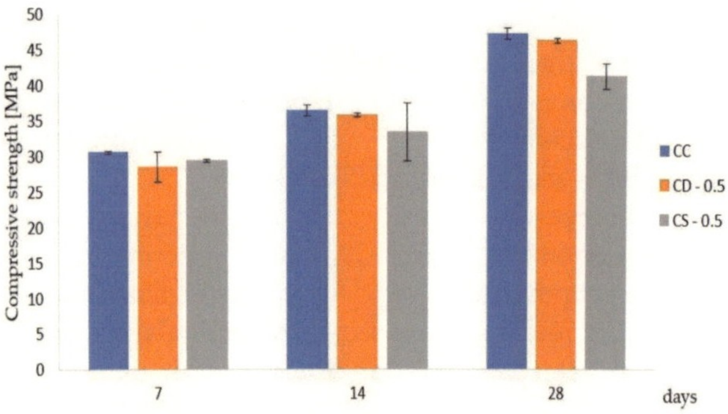

Figure 4.
Compressive strength results of concrete samples [29].

Although the limitation of microcracking in concretes was not specifically investigated, the addition of fibres is an accepted technique to control micro-cracking [88–90]. Moreover, Soto et al. [91] and Okeola et al. [92] demonstrated this using natural fibres. In particular, Araya-Letelier et al. [97] demonstrated the possibility of limiting the number and size of concrete microcracks using Chilean cements and natural fibre with 54% lower tensile strength than *Eucalyptus Globulus* bark fibres.

4.3.2 Incorporation of end-of-life tyre rubber into concrete

End-of-life tyres (ELT) represents a social and environmental problem due to the large accumulation in landfills. In fact, more than 1000 million tonnes of ELT are generated annually and more than 50% of that amount is destined to landfill or left as untreated garbage [98]. Moreover, this material produces concentrations of rats, larvae, mice and insects, increases the risk of fires difficult to extinguish [99]. This condition is aggravated by its lack of reutilisation. For instance, in Chile more than 140,000 tonnes of ELT were produced in 2019 and only 17% was recycled [100].

An attractive use of this waste is its incorporation in concrete mixes. Indeed, concrete with rubber improves properties such as energy absorption, due to its increased plasticity and ductility [101], which improves its impact behaviour and durability. However, the addition of rubber reduces the mechanical strength of the concrete [102, 103]. Therefore, to mitigate this effect, it is necessary to treat the rubber prior to its incorporation [104–106]. For instance, García et al. [107] evaluated the effects of three different treatments applied to ELT rubber before its incorporation in cement mortar samples. The treatments were hydration, oxidation-sulphonation and contact with hydrogen peroxide. The results indicate that it is possible to replace up to 5% of the fine aggregate weight with ELT rubber. Furthermore, the incorporation of hydrated ELT rubber proved to be the best treatment option from a technical, practical and economical point of view.

As the purpose is to produce a useful improvement in concrete properties and utilise waste efficiently, it is important to consider the geo-dependence of the concrete material [29]. This is particularly evident in samples made with cement containing local fly ash (Great Concepción, Chile). Indeed, this cement has a similar market cost to others national cements. However, the samples with ELT and fly ash cement are able to provide mortars that can satisfy the design requirements. Therefore, it is possible to reduce the amount of clinker while reusing a local industrial waste.

Furthermore, the adequate strength obtained in the experimental evaluation makes ELT rubber concrete an attractive alternative to improve the in-service performance of structures such as concrete crash barriers. Since rubber concrete is able to increase impact resistance and energy absorption [101], the durability of these safety elements is increased. Actually, studies indicate that concrete barriers with recycled rubber ELT have a service life 2.6 to 3.2 times longer than traditional barriers [108].

Due to the extensive road networks of the countries the benefits of this type of barrier represent not only a massive use of an industrial waste, but also bring economic, safety, and environmental benefits to the country or region where they can be applied. For instance, only in Chile the national road network roads exceed 85,000 kilometres [109].

Although the importance of geo-dependency, the results obtained in this research may be useful for other regions as well. For example, where it is possible to replace part of the cement with fly ash (produced or imported), or another cement substitute, producing similar results. If the substitute is a waste product such as fly ash, the contribution is not only technical and economic, but also it is a contribution to a sustainable development.

4.3.3 Biological self-healing concrete

A promising approach to mitigate durability problems and control microcracking of concrete is to incorporate biological elements that self-heal the material. This technique was originally developed by Dr. H. Jonkers at Delft University of Technology, who incorporate into the concrete a bacterium that has the ability to segregate calcium carbonate, sealing concrete microcracks [110]. This can represent an increment of 20% in the lifetime of a concrete structure [111]. If the same structure but with traditional concrete will need rehabilitation, and then more concrete, during its life to endure the same period of time than the self-healing concrete. Therefore, the amount of CO2 emitted into the environment decreases compared to structures built with conventional concrete. Indeed, Van Belleghem et al. [112] report that the lifetime of structures with self-repairing concrete can reduce the environmental impact by 56–75%.

Due to the biological nature of this technology and its interaction with the components of the concrete mix and the local environmental conditions, geodependence is relevant to the performance of the self-repair process. Indeed, if the ideal conditions of moisture, salinity and temperatures are not present, then the bacteria will not have sufficient stimulus to secrete calcium carbonate and the crack sealing will not be produced [113]. Similarly, cements with a higher amount of pozzolan decrease calcium carbonate precipitation which reduce the self-healing capacity of concrete [114].

In collaboration with Dr. Jonkers, bacterial growth, spore formation and production of concrete samples were developed with two types of Porland pozzolanic cements commonly used in Chile [115]. The *Bacillus pseudofirmus* bacteria solution was impregnated into expanded clays and incorporated into the concrete as a partial replacement of fine aggregate. These specimens were subjected to three different environments simulated in the laboratory: fresh water, salt water and 90% relative humidity, all at 20°C. The results at 28 days show some areas where the self-healing was 100%. Although in other cases the recovery was lower, about 90% of the specimens showed traces of calcium carbonate as a sign of the activation of the self-healing process (**Figure 5**).

The results confirm that biological self-healing of concrete can be produced with Chilean materials and regional environments simulated in the laboratory. In particular, the aggregates and cements used in the research were different from those used in the investigations of Jonkers [23, 116, 117].

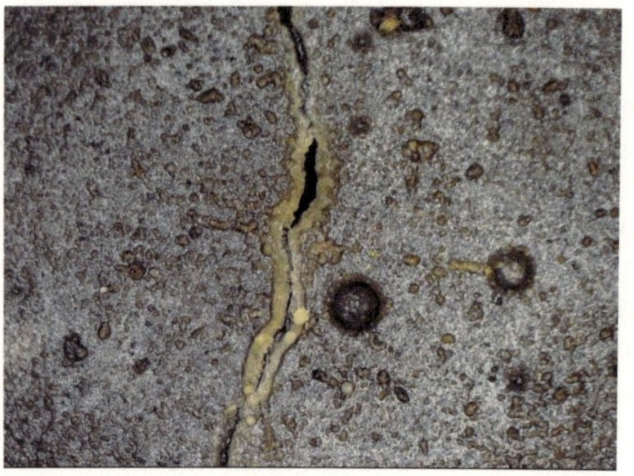

Figure 5.
Concrete sample with self-healing process activated [115].

4.4 Incorporating waste into the concrete as a mean of reducing the demand for natural aggregates

4.4.1 Cooper slag as a partial replacement of concrete fine aggregates

Copper slag (CS) is a vitreous substance generated in the copper smelting process. Globally, the copper industry produces approximately 24.6 million tonnes of slag [118]. In this regard, Chile is the world's largest copper producer [119, 120], generating 2.2 tonnes of CS for one tonne of copper produced. This creates significant accumulations of slag, negatively impacting the environment.

Considering the mechanical and chemical characteristics of CS, their use in concrete mixes as a partial replacement of aggregates has been evaluated. In fact, studies indicate that replacing up to 40% of fine aggregate in concrete samples results in higher or equivalent strengths compared to samples without slag [121–123]. Similar results were reported by Al-Jabri et al. [124] and Borkowsky [125] on mortar samples with 40% and 50% fine aggregate replacement, respectively. This percentage of replacement were experimentally verified by Pérez [126] for Chilean cement and aggregates. In addition, and in order to quantify the environmental benefits of this technique, a sustainability assessment based on the five-step method of Ashby et al. [30] was performed by Pérez [126] as well. The assessment estimates the percentage of CS that could be recycled, and then, the reduction of sand demand of cement mortars and concrete. Different scenarios were considered in the analysis that the copper smelters and the aggregate extraction plants are at similar distances from the concrete plants.

The results of the sustainability analysis indicate that in the most optimistic scenario, it is possible to reuse between 58% and 64% of the CS volume produced annually. This represents an annual saving of approximately 0.7 million tonnes of sand and an annual energy saving of 210409 GJ,1737905 kg CO2 eq, 11.89 kg Sb eq, 23549348 MJ ADP, 0.08 kg CDC-11 eq, 704.14 kg C2H4 eq, 12016 kg SO2 eq and 1563 kg PO4–3 eq. For instance, shows the benefits achieved for one of the cases evaluated (**Figure 6**).

These results are relevant since, in Chile, approximately 60% of building surfaces are made of concrete and 12% of masonry [127]. Therefore, there is a great opportunity to reuse this industrial waste.

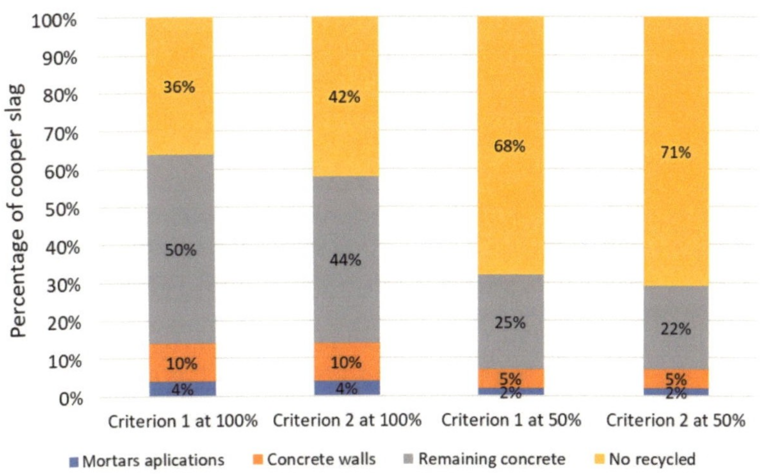

Figure 6.
Percentage of recycled CS [126].

Currently, the investigation continues at the concrete laboratory of Universidad de Concepción, with promising results in replacing more than 50% of the of the fine aggregate with CS.

4.4.2 Recycled plastics as a partial replacement of concrete fine aggregates

Various types of plastic waste have been incorporated into concrete to avoid direct contact with the environment [128–130]. In particular, in the Biobio region of Chile, the acrylonitrile butadiene styrene (ABS) is not being recycled and then its incorporation into concrete as a partial replacement of fine aggregate has been studied at the concrete laboratory of the Universidad de Concepcion (Biobio Region).

The ABS is a plastic used in the production of computer keys, appliance and power tool housings, plastic plug protectors and automotive parts such as dashboards and bumpers [131]. However, in Chile, the ABS accounts for only 2% of national plastic recycling [132], which can lead to serious waste disposal problems.

Preliminary results show that it is possible to replace up to 50% of the fine aggregate for applications as lean concrete, simple foundations, false floors and subfloors. However, optimal results were obtained when replacing 25% of the fine aggregate with ABS plastic.

Considering that concrete is a material with a high demand for natural aggregates, the strategy of using products with post-consumer recycled content, such as ABS plastic to replace sand in concrete, reduces the environmental impacts resulting from the initial stage of the concrete life cycle, extraction and processing of virgin materials.

5. Conclusions

As important as it is, sustainability related with the concrete material is more than reducing the amount of cement in the mixes. Actually, in this book chapter other approaches have been presented and analysed, starting with the phases of the concrete material life-cycle where the proposed alternatives present lower environmental impacts than the traditional ones.

Although the article mentions different examples of the approaches to develop more sustainable concrete alternatives, it goes more in details in the ones studied (or under study) at the concrete laboratory of the Universidad de Concepción. In this way, the article analyses how it is possible to reduce the demand of concrete, and therefore cement, by optimising the designs of concrete pavements. Another approach is the replacement of concrete aggregates by industrial waste such as copper slag or ABS plastic, which not only reduces the demand for the natural resource, but it contributes to solving a waste disposal problem. Finally, it is propose the incorporation, during the production stage, of natural fibres, ELT rubber or biological elements that can increase the concrete durability, which is a direct contribution to sustainability.

Hence, it is possible to develop more sustainable alternatives than the traditional ones applied in the concrete industry, which is especially relevant considering the massive use of the concrete material. However, for the development of viable alternatives, it is necessary to consider the geo-dependence of concrete. Indeed, the physical and chemical properties of the concrete components, together with the site conditions, can influence the concrete performance.

Acknowledgements

The authors express their gratitude to Dr. H.M. (Henk) Jonkers, Dr. F. (Francesco) Di Maio, Prof.dr.ir. M.J.G. (Sandra) Erkens, Dr. G. (Guang) Ye and PhD candidate Shizhe Zhang for the support received during the work developed in collaboration with Delf University of Technology. Additionally, special thanks are giving to the projects ANID PIA/Apoyo CCTE AFB170007 and VRID Multidisciplinario 219.091.051-M for funding received.

Conflict of interest

The authors declare no conflict of interest.

Author details

Mauricio Pradena* and Andrés César
Facultad de Ingeniería, Departamento de Ingeniería Civil, Universidad de Concepción, Concepción, Chile

*Address all correspondence to: mpradena@udec.cl

References

[1] Mehta PK, Monteiro PJM. Concrete: microstructure, properties, and materials. McGraw-Hill Education. 2014.

[2] Flatt RJ, Roussel N, Cheeseman CR. Concrete: An eco material that needs to be improved. J Eur Ceram Soc. 2012; 32(11): 2787-2798. https://doi.org/10.1016/j.jeurceramsoc.2011.11.012

[3] Marinković S, Malešev M, Ignjatović I. Life cycle assessment (LCA) of concrete made using recycled concrete or natural aggregates. In Eco-Efficient Construction and Building Materials. Cambridge: Woodhead Publishing; 2014. p. 239-266.

[4] Kline J, Barcelo L. Cement and CO2, a victim of success! In: Proceedings of the IEEE-IAS/PCA 54th Cement Industry Technical Conference; 14-17 May 2012; San Antonio. TX. USA. 2012. p. 1-14.

[5] Monteiro PJM, Miller SA., Horvath A. Towards sustainable concrete. Nat. Mater. 2017;16:698-699.

[6] Mindess S. Sustainability of concrete. Developments in the Formulation and Reinforcement of Concrete. 2nd ed. Vancuver: Woodhead Publishing; 2019. p. 3-17.

[7] Fernandez A, Leung Y. Technology Roadmap-Low-Carbon Transition in the Cement Industry. 2018. Available from: https://webstore.iea.org/technolo gyroadmap-low-carbontransition-in-the-cement-industry

[8] Lepech MD, Keoleian GA, Li VC, Qian S. Design of green engineered cementitious composites for pavement overlay applications. In: Proceedings of the 1st International Symposium on Life Cycle Civil Engineering; 10-14 June 2008; Varenna: Italy. 2008. P. 837-842.

[9] Potgieter JH. An Overview of Cement production: How "green" and sustainable is the industry? Environ. Manag. Sustain. Dev. 2012;1, 14.

[10] Uribe P. Explotación y venta de áridos en la comuna de Puerto Montt [thesis]. Valdivia: Universidad Austral de Chile; 2011. (In Spanish)

[11] García C, Pérez P. Alteraciones geomorfológicas recientes en los sistemas fluviales mediterráneos de la Península Ibérica. Síntomas y problemas de incisión en los cauces. Revista de geografía Norte Grande.2014;59;25-44. (In Spanish) http://dx.doi.org/10.4067/S0718-34022014000300003

[12] Opoku A, Ahmed V. Embracing sustainability practices in UK construction organizations. Built Environ. Proj. Asset Manag. 2014;4: 90-107. DOI:10.1108/bepam-02-2013-0001.

[13] World Commission on Environment and Development. Our Common Future: Report of the World Commission on Environment and Development. Oxford: Oxford University Press; 1987. 383p.

[14] Rosen MA. Engineering Sustainability: A Technical Approach to Sustainability. Sustainability. 2012;4: 2270-2292.

[15] Brunatti C, Fernandez L. El Pavimento de Hormigón. Un Camino Sustentable. In: Proceedings of the I Congreso Hormigón Premezclado de las Américas; 8-10 November 2010; Argentina: Mar del Plata. (In Spanish)

[16] Khatibinia M, Khosravi S. A hybrid approach based on an improved gravitational search algorithm and orthogonal crossover for optimal shape design of concrete gravity dams. Appl Soft Comput. 2014;16:223-233. doi:10.1016/j.asoc.2013.12.008.

[17] Kaveh A, Zakian P. Stability based optimum design of concrete gravity dam using CSS, CBO and ECBO algorithms. Iran Univ Sci Technol. 2015;5(4):419-431.

[18] Ferdowsi A, Hoseini SM, Farzin S, Faramarzpour M, Mousavi SF. Shape Optimization of Gravity Dams Using a Nature-Inspired Approach. Journal of Soft Computing in Civil Engineering. 2020;4(3):65-78.

[19] Meng Y, Ling TC, Mo KH. Recycling of wastes for value-added applications in concrete blocks: An overview. Resources, Conservation and Recycling. 2018;138:298-312.

[20] Prabhu GG, Hyun JH, Kim YY. Effects of foundry sand as a fine aggregate in concrete production. Construction and building materials. 2014;70:514-521.

[21] ACI. Measurement of Properties of Fiber Reinforced Concrete; ACI 544.2R-89; American Concrete Institute: Farmington Hills, MI, USA; 2009.

[22] Saravanan N, Buvaneshwari M. Experimental Investigation on Behaviour of Natural Fibre Concrete (Sisal Fibre). Cellulose. 2018; 54:66.

[23] Jonkers H, Thijssen A, Muyzer G, Copuroglu O, Schalangen E. Application of bacteria as self-healing agent for the development of sustainable concrete. Ecological Engineering. 2008;36: 230-235.

[24] Saleh H, El-Saied F, Salaheldin T, Hezo A. Macro-and nanomaterials for improvement of mechanical and physical properties of cement kiln dust-based composite materials. Journal of Cleaner Production. 2018; 204:532-541.

[25] Saleh et al. 2019 Saleh H, El-Sheikh S, Elshereafy E, Essa A. Mechanical and physical characterization of cement reinforced by iron slag and titanate nanofibers to produce advanced containment for radioactive waste. Construction and Building Materials. 2019; 200: 135-145.

[26] Ninčević K, Boumakis I, Marcon M, Wan-Wendner, R. Aggregate effect on concrete cone capacity. Engineering Structures. 2019;191:358-369.

[27] Mindess S, Young JF, Darwin D. Concrete.2nd ed. Upper Saddle River, NJ: Prentice-Hall. 2003

[28] Zayed A. Effects of portland cement particle size on heat of hydration (No. BDK84 977-13). Florida: Dept. of Transportation. Research Center; 2013.

[29] Mansilla C, Pradena M, Fuentealba C, César A. Evaluation of Mechanical Properties of Concrete Reinforced with Eucalyptus globulus Bark Fibres. Sustainabilty. 2020;12:10026.

[30] Ashby M, Coulter P, Ball, N. The CES EduPack Eco Audit Tool – A White Paper. Granta Design Ltd. 2011; 2:23.

[31] Hassan HS, Abdel-Gawwad HA, Vásquez-García SR, Israde-Alcántara I, Flores-Ramirez N, Rico JL, Mohammed MS. Cleaner production of one-part white geopolymer cement using pre-treated wood biomass ash and diatomite. Journal of Cleaner Production. 2019;209: 1420-1428.

[32] Maschio S, Tonello G, Piani L, Furlani E. Fly and bottom ashes from biomass combustion as cement replacing components in mortars production: Rheological behaviour of the pastes and materials compression strength. Chemosphere. 2011;85(4): 666-671.

[33] Rissanen J, Ohenoja K, Kinnunen P, Illikainen M. Partial replacement of portland-composite cement by fluidized bed combustion fly ash. Journal of

Materials in Civil Engineering. 2017;29(8):04017061.

[34] IS:1489(Part1), Product Manual For Portland Pozzolana Cement –Fly Ash based according to IS 1489 (Part 1): 2015, Bur. Indian Stand. Dehli. 2018.1489.

[35] Dhawan A, Gupta N, Goyal R, Saxena KK. Evaluation of mechanical properties of concrete manufactured with fly ash, bagasse ash and banana fibre. Materials Today. 2021;44:17-22.

[36] Fořt J, Šál J, Ševčík R, Doleželová M, Keppert M, Jerman M, Černý R. Biomass fly ash as an alternative to coal fly ash in blended cements: Functional aspects. Construction and Building Materials. 2021;271: 121544.

[37] Kupwade-Patil K, Palkovic SD, Bumajdad A, Soriano C, Büyüköztürk O. Use of silica fume and natural volcanic ash as a replacement to Portland cement: Micro and pore structural investigation using NMR, XRD, FTIR and X-ray microtomography. Construction and Building Materials. 2018;158:574-590.

[38] Jagan S, Neelakantan TR. Effect of silica fume on the hardened and durability properties of concrete. International Review of Applied Sciences and Engineering. 2021; 12(1):44-49.

[39] Sun J, Wang Z, Chen Z. Hydration mechanism of composite binders containing blast furnace ferronickel slag at different curing temperatures. Journal of Thermal Analysis and Calorimetry. 2018; 131(3): 2291-2301.

[40] Abdel Gawwad HA, Abd El-Aleem S, Faried AS. Influence of nano-silica and-metakaolin on the hydration characteristics and microstructure of air-cooled slag-blended cement mortar. Geosystem Engineering. 2017;20(5):276-285.

[41] Mo KH, Mohd Anor FA, Alengaram UJ, Jumaat MZ, Rao KJ. Properties of metakaolin-blended oil palm shell lightweight concrete. European Journal of Environmental and Civil Engineering. 2018; 22(7):852-868.

[42] Pan SC, Tseng DH, Lee C. Use of sewage sludge ash as fine aggregate and pozzolan in portland cement mortar. Journal of Solid Waste Technology and Management. 2002;28(3): 121-130.

[43] Lin KL, Chang WC, Lin DF, Luo HL, Tsai MC. Effects of nano-SiO2 and different ash particle sizes on sludge ash–cement mortar. Journal of environmental management. 2008;88(4):708-714.

[44] Alex J, Dhanalakshmi J. Ambedkar B. Experimental investigation on rice husk ash as cement replacement on concrete production. Construction and Building Materials. 2016;127:353-362.

[45] Puertas F, Varga C, Palacios M, Pellerin B, Eychenne-Baron C, Babayan D, Elkhadiri I. Alkali-activation of slag cements: Activation process, microstructure and mechanical properties. In: Proceedings of 13th International Congress on the Chemistry of Cement; 3-8 July 2011; Madrid. p. 1-7.

[46] Garcia-Lodeiro I, Fernandez-Jimenez A, Palomo A. Hydration kinetics in hybrid binders: Early reaction stages. Cement and Concrete Composites. 2013;39: 82-92.

[47] Angulo-Ramírez DE, de Gutiérrez RM, Puertas F. Alkali-activated Portland blast-furnace slag cement: Mechanical properties and hydration. Construction and Building Materials. 2017;140:119-128.

[48] Wianglor K, Sinthupinyo S, Piyaworapaiboon M, Chaipanich A. Effect of alkali-activated metakaolin

cement on compressive strength of mortars. Applied Clay Science. 2017;141:272-279.

[49] Kovalchuk G, Fernandez-Jimenez A, Palomo A. Activacion alcalina de cenizas volantes. Relación entre el desarrollo mecánico resistente y la composición química de la ceniza. Materiales de Construcción. 2008; 58: 35-52.

[50] Torres-Carrasco M, Puertas F. La activación alcalina de diferentes aluminosilicatos como una alternativa al Cemento Portland: cementos activados alcalinamente o geopolímeros. Revista ingeniería de construcción. 2017;32(2): 5-12.

[51] Hardjito D, Wallah SE, Sumajouw DM, Rangan BV. On the Development of Fly Ash–Based Geopolymer Concrete. ACI Materials Journal. 2004;101(6):467-472.

[52] Tempest B, Sanusi O, Gergely J, Ogunro V, Weggel, D. Compressive Strength and Embodied Energy Optimization of Fly Ash Based Geopolymer Concrete. In: Proceedings World of Coal Ash Conference; 7-9 May 2009; Lexington, KY: p.1-17

[53] Deepika R, Suribabu CR. Optimal design of gravity dam using differential evolution algorithm. Iran Univ Sci Technol. 2015;5(3):255-266.

[54] Hashemian FK. Structural Behaviour and Optimization of Moment-Shaped Reinforced Concrete Beams. Ann Arbor: University of Manitoba (Canada); 2012.

[55] López Vidal A, Bartolomé Muñoz, C. Fachadas de hormigón arquitectónico. Industrialización, eficiencia y estética. Alzada. 2019;119:52-57.

[56] Bartolomé C, Alarcón A, Tenorio Ríos JA, Bermejo E. New materials to increase the thermal mass of existing buildings for its energy rehabilitation. 2020

[57] Whiteley DL, Goethert K, Goodwin F, Golter HP, Kennedy J, Watternburg T, Meyer Jessi, et al. Sustainability for Repairing and Maintaining Concrete and Masonry Buildings. ICRI Committee 160. Rosemont, IL, USA; 2014.

[58] Otieno M, Alexander M, Beushausen HD. Corrosion in cracked and uncracked concrete–influence of crack width, concrete quality and crack reopening. Magazine of Concrete Research. 2010;62(6): 393-404.

[59] Snoeck D. Superabsorbent polymers to seal and heal cracks in cementitious materials. RILEM Technical Letters. 2018;30: 32-38.

[60] Paul SC, van Zijl GPAG., Šavija B. Effect of Fibers on Durability of Concrete: A Practical Review. Materials. 2020;13:4562.

[61] Ann KY, Ahn JH, Ryou JS. The importance of chloride content at the concrete surface in assessing the time to corrosion of steel in concrete structures. Construction and Building Materials. 2009;23(1): 239-245.

[62] Castañeda OH, Escobedo CJM. Durabilidad e infraestructura: retos e impacto socioeconómico. Ingeniería. Investigación y Tecnología. 2006;7(1):57-70.

[63] Berrocal CG, Löfgren I, Lundgren K, Tang L. Corrosion initiation in cracked fibre reinforced concrete: influence of crack width, fibre type and loading conditions. Corrosion Science. 2015;98: 128-139.

[64] Otieno M, Beushausen H, Alexander M. Resistivity-based chloride-induced corrosion rate prediction models and hypothetical

framework for interpretation of resistivity measurements in cracked RC structures. Materials and Structures. 2016;49(6): 2349-2366.

[65] Meredith J, Ebsworth R, Coles SR, Wood BM, Kirwan K. Natural fibre composite energy absorption structures. Compos. Sci. Technol. 2012;72:211-217.

[66] Andiç-Çakir Ö, Sarikanat M, Tüfekçi HB, Demirci C, Erdogan ÜH. Physical and mechanical properties of randomly oriented coir fiber–cementitious composites. Compos. B Eng. 2014;61:49-54.

[67] Ardanuy M, Claramunt J, Toledo Filho RD. Cellulosic fiber reinforced cement-based composites: A review of recent research. Constr. Build. Mater. 2015;79:115-128.

[68] Dash MK, Patro SK, Rath AK. Sustainable use of industrial-waste as partial replacement of fine aggregate for preparation of concrete–A review. International Journal of Sustainable Built Environment. 2016;5(2):484-516.

[69] Miller SA, Horvath A, Monteiro PJ. Impacts of booming concrete production on water resources worldwide. Nature Sustainability. 2018;1(1):69-76.

[70] U.S. Department of Transportation Federal Highway Administration. Transportation Applications of Recycled Concrete Aggregate—FHWA State of the Practice National Review. Washington, DC, USA; 2004. 47p.

[71] DiarioUChile, Los Peligros de la Extracción Masiva de Áridos [Internet]. 2015. Available from: https://radio.uchile.cl/2015/02/08/los-peligros-de-la-extraccion-masiva-de-aridos/ [Accessed: 2021-06-14]

[72] Rajan MS. Study on strength properties of concrete by partially replacement of sand by steel slag. International Journal of Engineering, Science and Technology. 2014;1(6): 96-99.

[73] Leema A, Suganya P. Performance of copper slag on strength and durability properties as partial replacement of fine aggregate in concrete. IJETAE. 2015;5(1):434-437.

[74] Thorneycroft J, Orr J, Savoikar P, Ball RJ. Performance of structural concrete with recycled plastic waste as a partial replacement for sand. Construction and Building Materials. 2018;161:63-69.

[75] Plati C. Sustainability factors in pavement materials, design, and preservation strategies: A literature review. Constr. Build. Mater. 2019;211: 539-555.

[76] Van Dam T, Taylor P, Fick G, Gress D, VanGeem M, Lorenz E. Sustainable Concrete Pavements: A Manual of Practice. Ames: National Concrete Pavement Technology Center, Iowa State University; 2012.

[77] Covarrubias TJP, Covarrubias JP. TCP design for thin concrete pavements. In: Proceedings of the 9th International Conference on Concrete Pavements; 17-21 August 2008; San Francisco, CA, USA p. 905-917.

[78] Covarrubias JP. Design of concrete pavement with optimized slab geometry. Rev. Ing. Constr. 2012; 27: 181-197.

[79] Mohod MV, Kadam KN. A comparative study on rigid and flexible pavement: A review. IOSR JMCE. 2016;13:84-88.

[80] Skrzypczak I, Radwański W, Pytlowany T. Durability vs technical—The usage properties of road pavements. E3S Web Conf. 2018;45:00082.

[81] Chávez O. Losas Cortas: Una experiencia en Nicaragua. In: Proceedings

of the VII Congreso Nacional de Ingeniería Civil; 17-18 October 2013; Managua, Nicaragua. p. 17-18

[82] Salsilli R, Wahr C, Delgadillo R, Huerta J, Sepúlveda P. Field performance of concrete pavements with short slabs and design procedure calibrated for Chilean conditions. Int. J. Pavement Eng. 2014;16:363-379. DOI:10. 1080/10298436.2014.943129.

[83] Pradena M, Houben L. Load Transfer-Crack Width Relation of Non-Dowelled Jointed Plain Concrete Short Slabs. Balt. J. Road Bridge Eng. 2018;13:40-45. DOI:10.3846/ bjrbe.2018.388.

[84] Roesler JR, Cervantes VG, Amirkhanian A. Accelerated performance testing of concrete pavement with short slabs. Int. J. Pavement Eng. 2012;13:494-507.

[85] César A, Pradena M. Sustainable Engineering: Load Transfer Characterization for the Structural Design of Thinner Concrete Pavements. Sustainability. 2020;12(21):9153.

[86] ISO. ISO 14040: Environmental management—Life cycle assessment—Principles and Framework; ISO: Geneva, Switzerland, 2006.

[87] ISO. ISO 14044: Environmental management—Life cycle assessment—Requirements and Guidelines; ISO: Geneva, Switzerland, 2006.

[88] Mihashi H, Ahmed SFU, Kobayakawa A. Corrosion of reinforcing steel in fiber reinforced cementitious composites. Journal of Advanced Concrete Technology. 2011;9(2):159-167.

[89] Luković M, Hordijk DA, Huang Z, Schlangen E. Strain hardening cementitious composite (SHCC) for crack width control in reinforced concrete beams. Heron. 2019; 64(1/2):181.

[90] Bertelsen IMG, Ottosen LM, Fischer G. (2020): Influence of fibre characteristics on plastic shrinkage cracking in cement-based materials: A review. Construction and Building Materials. 2020;230:116769.

[91] Soto II, Ramalho MA, Izquierdo OS. Post-cracking behavior of blocks, prisms, and small concrete walls reinforced with plant fiber. Revista IBRACON de Estruturas e Materiais. 2013;(6):598-612.

[92] Okeola AA, Abuodha SO, Mwero J. The effect of specimen shape on the mechanical properties of sisal fiber-reinforced Concrete. Open Civil Engineering Journal. 2018;12: 368-382.

[93] Pehanich J, Blankenhorn P, Silsbee M. Wood fibre surface treatment level effects on selected mechanical properties of wood fibre-cement composites. Cement and Concrete Research. 2004;34:59-65.

[94] Yalley PP, Kwan ASK. Use of coconut fibre as an enhancement of concrete. Journal of Engineering Technology. 2009;3:54-73.

[95] Zakaria M, Ahmed M, Hoque MM, Islam S. Scope of using jute fiber for the reinforcement of concrete material. Textiles and Clothing Sustainability. 2017;2(1):1-10.

[96] Bown HE, Lasserre JP. An air-drying model for piled logs of Eucalyptus globulus and Eucalyptus Nitens in Chile. N. Z. J. Forestry Sci. 2015;45(1):1-9.

[97] Araya-Letelier G, Antico FC, Carrasco M, Rojas P, García-Herrera CM. Effectiveness of new natural fibers on damage-mechanical performance of mortar. Construction and Building Materials. 2017;152: 672-682.

[98] Thai QB, Le DK, Do NH, Le PK, Phan-Thien N, Wee CY, Duong HM.

Advanced aerogels from waste tire fibers for oil spill-cleaning applications. J. Environ. Chem. Eng. 2020;8:104016.

[99] Kashani A, Ngo TD, Hemachandra P, Hajimohammadi A. Effects of surface treatments of recycled tyre crumb on cement-rubber bonding in concrete composite foam. Constr. Build. Mater. 2018;171:467-473.

[100] Gobierno Regional de Arica y Parinacota. [Internet]. 2018. Available from: https://www.goreayp.cl/index.php/noticias/1962-nueva-ley-paracombatir-140-mil-toneladas-de-neumaticos-en-desuso (Accessed: 20020-12-1).

[101] Atahan AO, Yücel AÖ. Crumb rubber in concrete: static and dynamic evaluation. Construction and Building Materials. 2012;36:617-622.

[102] Youssf O, Mills JE, Benn T, Zhuge Y, Ma X, Roychand R, Gravina R. Development of Crumb Rubber Concrete for Practical Application in the Residential Construction Sector–Design and Processing. Constr. Build. Mater. 2020;260:119813.

[103] Rashid K, Yazdanbakhsh A, Rehman MU. Sustainable selection of the concrete incorporating recycled tire aggregate to be used as medium to low strength material. J. Clean. Prod. 2019;224:396-410.

[104] Shatanawi KM, Biro S, Naser M, Amirkhanian SN. Improving the rheological properties of crumb rubber modified binder using hydrogen peroxide. Road Mater. Pavement Des. 2013;14:723-734.

[105] Mohammadi I, Khabbaz H, Vessalas K. In-depth assessment of Crumb Rubber Concrete (CRC) prepared by water-soaking treatment method for rigid pavements. Constr. Build. Mater. 2014;71: 456-471.

[106] He L, Ma Y, Liu Q, Mu Y. Surface modification of crumb rubber and its influence on the mechanical properties of rubbercement concrete. Constr. Build. Mater. 2016;120:403-407.

[107] García E, Villa B, Pradena M, Urbano B, Campos-Requena VH, Medina C, et al. Experimental Evaluation of Cement Mortars with End-of-Life Tyres Exposed to Different Surface Treatments. Crystals. 2021;11(5):552.

[108] Raj A, Arshad PU, Nagarajan P, Shashikala AP. Experimental investigation on the fracture behaviour of polypropylene fibre-reinforced rubcrete. In Structural Integrity Assessment. 2020 (pp. 335-345). Springer, Singapore.

[109] MOP. Red Vial Nacional Dimensionamiento y características. Chile: Dirección de Vialidad. Ministerio de Obras Públicas. 2020. Available from: http://www.vialidad.cl/areasdevialidad/gestionvial/Documents/RedVial Nacional2019.pdf

[110] Jonkers H. (2007) Self healing concrete: A biological Approach. Self healing materials. Springer, Dordrecht. 2007;195-204.

[111] Jonkers, H. New Materials: Biulding with Living Materials - Henk Jonkers, researcher at TU Delft [Web streaming video]. France: Hello Tomorrow; 2017 [cite 2021 May 21]. Available from: https://www.youtube.com/watch?v=xi6RJ4DpbVY

[112] Van Belleghem B, Van den Heede P, Van Tittelboom K, De Belie N. Quantification of the service life externsion and environmental benefit of chloride exposed self-healing concrete. Materials. 2017;10(1): 5

[113] Parraguez A. Estudio sobre el rendimiento de bacterias como agente de auto-reparación en el hormigón bajo

diferentes condiciones de temperatura y tipo de cemento [thesis]. Valparaíso: Universidad Técnica Federico Santa María; 2018.

[114] Becker, E. Cemento Portland–Características y Recomendaciones de Uso. Publicación de Loma Negra CIASA. 2000.

[115] Ramírez F. Evaluación de la autocuración biológica de microgrietas en hormigones bajo condiciones simuladas de laboratorio [thesis]. Concepción: Universidad de Concepción; 2020.

[116] Wiktor V, Jonkers H. Quantification of crack-healing in novel bacteria-based self-healing concrete. Cement & concrete composites. 2011;33:763-770.

[117] Jonkers H. Bacteria-based self-healing concrete. Heron. 2011;56(1/2):1-12.

[118] Gorai B, Jana RK. Characteristics and utilization of copper slag – a review. Conservation and Recycling. 2003; 39(4):299-313.

[119] Comisión Chilena del Cobre. Sección Estadísticas. Producción de cobre de mina mundial y Chile. COCHILCO, Ministerio de Minería. Chile; 2018.

[120] Minería Chilena. Participación de Chile en el mercado mundial de cobre cayó a 28% en 2019 y los rivales recortan terreno [Internet].2020. Available from: https://www.mch.cl/2020/03/16/participacion-de-chile-en-el-mercado-mundial-del-cobre-cayo-a-28-en-2019-y-los-rivales-recortan-terreno/# [Accessed: 2021-6-15]

[121] Cendoya, P. Efecto en la resistencia de las escorias de fundición de cobre como agregad fino en el comportamiento resistente del hormigón. Ingeniare. Revista chilena de ingeniería. 2009;17(1):85-94.

[122] Brindha D, Nagan S. Utilization of copper slag as a partial replacement of fine aggregate in concrete. International Journal of Earth Sciences and Engineering. 2010;3(4):579-585.

[123] Thomas J, Thaickavil N, Abraham MP. Copper or ferrous slag as substitutes for fine aggregates in concrete. Advances in Concrete Construction. 2018;6:545-560.

[124] Al-Jabri KS, Al-Saidy H, Taha R. Effect of copper slag as a fine aggregate on the properties of cement mortars and concrete. Construction Building Materials. 2011;25:933-938.

[125] Borkowsky, A. Estudio de la falla de corte de morteros de pega con escorias de fundición de cobre [thesis]. Concepción: Universidad de Concepción; 2018.

[126] Pérez M. Análisis de la utilización de escoria de fundición de cobre como alternativa sustentable de reemplazo parcial de árido fino en hormigones [thesis]. Concepción: Universidad de Concepción; 2021.

[127] Instituto del Cemento y del Hormigón de Chile. Material muros, superficie total y participación. [Internet]. 2019 https://ich.cl/estadisticas/material-muros-superficie-total-y-participacion/ [Accessed: 2020-9-23]

[128] Marzouk OY, Dheilly RM, Queneudec M. Valorisation of post-consumer plastic waste in cementitious concrete composites. Waste Manage. 2007;27(2):310-318.

[129] Choi YW, Moon DJ, Kim YJ, Lachemi M. Characteristics of mortar and concrete containing fine aggregate manufactured from recycled waste polyethylene terephthalate bottles. Constr Build Mater 2009;23: 2829-2835.

[130] Akçaözoğlu S, Atiş CD, Akçaözoğlu K. An investigation on the use of shredded waste PET bottles as aggregate in lightweight concrete. Waste management. 2010;30(2):285-290.

[131] Pérez Fernández A. Estudio de la cinética de degradación térmica de un acrilonitrilo butadieno estireno (ABS) reciclado mediante termogravimetría [thesis]. Valencia: Universitat Politécnica de Valencia; 2020.

[132] ASIPLA. Estudio sobre el Reciclaje de Plásticos en Chile [Internet]. 2019. Available from: http://www.asipla.cl/wp-content/uploads/2019/04/190328-Estudio-sobre-Reciclaje-de-Pl%C3%A1sticos-en-Chile-Resumen-Ejecutivo.pdf [Accessed: 2021-4-15]

Chapter 5

Bearing Capacity of Concrete Filled Steel Tube Columns

Anatoly Krishan

Abstract

Concrete filled steel tubes columns of circular cross section (CFST) have significant constructive, technological, economic advantages. Therefore, CFST are increasingly used in construction practice. Due to the complex nature of CFST load resistance, regulations of the Europe, Australia, Brazil, India, Canada, China, the USA, Japan, and of a number of other countries recommend using empirical formulas for calculating their bearing capacity. Despite the large number of the experiments, serving as a basis for these formulas, they do not always allow to obtain valid results. Besides, these methods, as a rule, do not allow the calculations of compressed CFST elements, which have any differences from a "classical" design, for example, the presence of a high-strength rod and (or) spiral reinforcement, various types of concrete, the effect of preliminary lateral compression of a concrete core, etc. The purpose of this monograph is to propose the method of deformation calculation of the bearing capacity of compressed CFST elements under short-term load action based on the phenomenological approach and the theoretical positions of reinforced concrete mechanics.

Keywords: concrete filled steel tube columns, spiral reinforcement, method of deformation calculation, strength, deformation, flexibility

1. Introduction

Concrete filled steel tubes columns (CFST) are composite structures. They feature a variety of advantages. CFST have significant constructive, technological, economic advantages and at the same time an architecturally expressive appearance [1–5]. Such obvious CFST advantages as decreased labor consumption of their production due to lack of forms and reinforcement cages and high speed of building erection are quite attractive for construction specialists. Besides, mechanical features of a steel shell and a concrete core combine quite rationally in these columns. The strong steel shell serves as a reliable frame for the concrete core ensuring good volumetric load conditions for it. Due to this, concrete strength of columns with circular cross-section increases 1.8÷2.5 times in average. Concrete, in its turn, protects the walls of the steel shell from loss of stability and corrosion from inside. As a result, concrete and steel mutually increase load-carrying ability of each other and that of the whole element.

In case of emergency (explosions, earthquakes, etc.), another important feature of such columns, high survivability, comes to the fore. It is ensured by high deformability of the concrete core, which, together with its high strength, ensures absorption of large amounts of energy during strength resistance of the construction. Therefore, CFST of circular cross-section are increasingly used in construction practice.

The high strength and deformability of the concrete core ensure its main advantages, especially for short centrally loaded circular cross-section concrete-filled tubular elements. Due to the complicated nature of CFST load resistance, regulations of the Europe, Australia, Brazil, India, Canada, China, the USA, Japan, and a number of other countries recommend using empirical formulas to calculate their bearing capacity.

Despite the large number of the experiments serving as a base for these formulas they do not always allow to obtain valid results [6, 7]. They have significant limitations in the field of application. They were obtained either from the results of specific laboratory sample testing, or due to statistical processing of the relevant data. First, these formulas are valid only for normal concrete. They give unreliable results for the columns from other types of concrete (for example, fine-grained ones). Secondly, these methods, as a rule, do not allow the calculations of eccentrically compressed concrete filled steel tube elements, which have any differences from a "classical" design, for example, the presence of a high-strength rod [8, 9] and (or) spiral reinforcement [10–12], the application of various types of concrete [13], the effect of preliminary lateral reduction in a concrete core [14], etc.

According to the results of researches carried out by many scientists, the most reliable calculations of the strength of CFST columns can be performed based on the recommendations of the EN 1992-1-1 standard. Moreover, a simplified method is often used in the calculations. But it is based on empirical formulas and is very limited in scope. It is proposed to consider the general case of calculation as well. For its implementation, the following assumptions are made:

- internal forces are determined by elasto-plastic analysis;
- plane sections may be assumed to remain plane;
- contact strength between steel and concrete components must be maintained up to column failure;
- the tensile strength of concrete is neglected.

Design of column structural stability should take into account second-order effects including residual stresses, yielding of structural steel and of reinforcement, local instability, cracking of concrete, creep and shrinkage of concrete, geometrical imperfections.

However, there are no specific methods for practical implementation of such a calculation.

The purpose of this monograph is to propose the method of deformation calculation of the bearing capacity of compressed CFST under short-term load action based on the phenomenological approach.

2. Basic design provisions

2.1 Initial provisions

Initially, the diameter d and wall thickness δ of the tube should be assigned for CFST. Taking into account the research results [7] for columns of circular cross-section, it is recommended to use the following restrictions:

$$20\sqrt{\frac{235}{f_y}} \leq \frac{d}{\delta} \leq 150\frac{235}{f_y}, \tag{1}$$

where f_y is a yield stress of the steel shell, MPa.

For monolithic columns, the possibility of loss of stability of the tube wall at the stage of installation of the supporting structures of the frame should be taken into account. The steel tube can be used as a supporting structure for several overlying floors even before it is filled with concrete, which significantly speeds up the process of constructing a building. In this case, local buckling is impossible when

$$\frac{d}{\delta} \leq 85\sqrt{\frac{235}{f_y}}. \tag{2}$$

If condition (2) is not met, it is necessary to check the stability of the tube walls under the action of corresponding loads. For this purpose, for example, the recommendations of European norm procedure (EN 1993-1-1 Steel Design) can be used.

For a short centrally loaded CFST column, the cross-sectional strength is usually determined. Most researchers use a fairly simple formula for this

$$N = f_{cc}A + \sigma_{pz}A_p, \tag{3}$$

where f_{cc} is strength of volumetrically loaded concrete core;

σ_{pz} is axial direction compression in the steel shell in CFST limit state;

A and A_p are cross-section areas of the concrete core and the steel shell.

Thus, in order to calculate the CFST strength, it is necessary to know the values of the strength of the volumetrically loaded concrete core and the compression in the steel shell. Various approaches and relationships for determining f_{cc} and σ_{pz} are recommended. They are reviewed below.

2.2 Known approaches for determining the strength of a concrete core

Compression strength is a very important mechanical attribute of CFST concrete core. In the limiting state centrally loaded circular section column, concrete is in the conditions of three-axis compression by axial direction strain σ_{cz} and transverse strain σ_{cr}.

A quite simple relationship, being in fact the Mohr-Coulomb strength condition, is most often used in calculations for such conditions

$$f_{cc} = f_c + k\sigma_{cr}, \tag{4}$$

where f_c is concrete unconfined compression strength;

k is coefficient of lateral pressure.

Considering experiments, the value of the k coefficient is usually taken as constant in this formula: k = 4,1 or k = 4,0.

Though the Eq. (4) was recommended by American researches F. Richard, A. Brandtzæg and R. Brown as far back as in 1929, it is currently used by many researches, including for designing columns with different types of confinement reinforcement. The relationships to determine the volumetrically loaded concrete recommended by regulations in many countries have been obtained based on this very formula. However, the gained new experimental materials evidence that the Eq. (4) does not always allow to get a valid result.

This is caused by many reasons. One of them is inaccuracies in determination of lateral strain σ_{cr}. The second reason is ignoring the scale factor. Since CFST frequently have significant cross-sectional dimensions (630 ... 1000 mm and more for high buildings), this factor shall be considered. The research [4] devoted to a review of a government program of concrete-filled tubes research carried out in the end of the 20th century in Japan introduces the following relationship

$$f_{cc} = \gamma_c f_c + k\sigma_{cr}, \tag{5}$$

in which γ_c scale is factor coefficient determined by the formula

$$\gamma_c = 1,67 \cdot d_c^{-0,112} \geq 0,85, \tag{6}$$

where d_c is concrete core diameter in mm.

A similar dependence was proposed in [15].

Regarding such approach as conceptually correct, it is worth mentioning a quite limited range of CFST cross section diameters, where usage of relationships (6) allows to obtain a result acceptable for practical purposes. According to this formula, first, $\gamma_c \approx 1$ when the concrete diameter is 100 mm, and a $\gamma_c \approx 0.95$ when d_c = 150 mm. In most countries, square-sided cube test pieces or cylinders with a cross-sectional diameter of 150 mm are considered reference concrete. In this case, the regulations provide that $\gamma_c = 1.05$ when d_c = 100 mm. Secondly, one has to take $\gamma_c = 0.85$ already when the cross-section diameter exceeds 300 mm, which does not correspond to experimental data of researches of large scaled samples with cross section diameters between 630 and 1020 mm.

Considering the results of the research [16], the coefficient γ_c is recommended to be determined by the formula

$$\gamma_c = 0,75 + 0,25\left(\frac{d_0}{d_c}\right)^{0,5}, \tag{7}$$

where d_o is reference cylinder diameter taken equal to 150 mm.

This formula does not need any limitations in a quite wide range of d_c = 100 to 3000 mm, which is convenient for practical calculations.

Another reason of the results obtained by the Eq. (4) not always corresponding to experimental data is the value of the coefficient of lateral pressure k = 4.1 taken as constant here. The research [17] shows theoretically that this value is variable. Our researches [16] found that the lateral pressure σ_{cr} reaches sometimes a value of 10÷15 MPa and more for CFST concrete cores before concrete destruction. Meanwhile, the values of the coefficient of lateral pressure can be within a range k =2,5÷7. Therefore, it is obvious that even insignificant inaccuracies in determination of k frequently lead to significant errors in determination of concrete core strength f_{cc} and load-carrying ability of a designed element.

Some of researches recommend considering this point. For example, in the research [18] it was correctly mentioned that, other factors being equal, the value of the coefficient of lateral pressure decreases while this pressure increases. A formula is recommended for its determination

$$k = 6,7(\sigma_{cr})^{-0,17}. \tag{8}$$

However, recently a formula of J. Mander has been used more frequently than others [19].

$$\frac{f_{cc}}{f_c} = 2,254\sqrt{1 + 7,94\frac{\sigma_{cr}}{f_c}} - 2\frac{\sigma_{cr}}{f_c} - 1,254. \tag{9}$$

This formula was received based on the results of statistical processing of a large amount of experimental data and is usable for not only medium- but also high-strength concrete with f_c of up to 120 MPa.

However, two main disadvantages of the Eq. (9) should be mentioned. First, lateral pressure σ_{cr} shall be known in CFST limit state to use it. As previously noted, this pressure is unknown when the load-carrying ability of such columns is calculated. Experiments with 180 samples of concrete-filled tubular elements [16] showed that σ_{cr} depends on geometry and design parameters of a designed column and may vary in wide limits. In addition, the relationship (9) is correct only for normal concrete. E.g. it is well known that fine grain concrete resists volumetric compression somewhat worse [17]. That is why other relationships shall be obtained for other concrete types, which causes certain inconveniences in calculations.

Processing of a number of experimental data evidences the existence of a stable relationship between σ_{cr} and a constructive coefficient of concrete-filled tubes ξ determined with use the formula

$$\xi = \frac{f_y A_p}{f_c A}. \tag{10}$$

The appropriate formulas are used in Chinese Technical Code for CFST structures (GB50936–2014).

2.3 State of stress in steel tube

Two methods to assess state of stress in a steel shell are known. The first one hypothesizes that a steel tube acts only transversely in limit state. In this case, the axial direction compression in the steel shell σ_{pz} is equal to zero. Then hoop stress determining the value of the lateral pressure in concrete reaches the yield stress of steel $\sigma_{p\tau} = f_y$. However, in general, it does not correspond to the real state of stress in a steel shell. Most researchers believe that the value of stress σ_{pz} depends on geometry and design parameters of CFST.

In the limiting state, the stress intensity in the steel shell reaches the yield point. During the central compression of a short CFST element, the steel shell experiences a compression-tension-compression stress state. Radial compressive stresses in the wall of steel tubes with $d/\delta \geq 40$ are small and they are usually neglected. Then the plane stress state "compression-tension" is considered for the tube. For this case, the Hencky-Mises yield criterion is written as follows:

$$\sigma_{pz}^2 + \sigma_{p\tau}^2 - \sigma_{pz}\sigma_{p\tau} = f_y^2, \tag{11}$$

where $\sigma_{p\tau}$ is the steel tube hoop stress in CFST limit state.

Then the stress σ_{pz} can be calculated using the formula

$$\sigma_{pz} = \sqrt{f_y^2 - 0,75\sigma_{p\tau}^2} - 0,5|\sigma_{p\tau}|. \tag{12}$$

Let us mention that the Eq. (12) is correct for thin-shell tubes when $d/\delta \geq 40$. These very tubes are generally used as steel shells for CFST.

The hoop stresses averaged by thickness in the steel shell for thin-shell tubes can be expressed through the lateral pressure by the following relationship with accuracy sufficient for practical calculations

$$\sigma_{p\tau} = -2\sigma_{cr}\frac{A}{A_p}. \tag{13}$$

Consequently, the axial direction compression in the steel shell depend on its yield stress f_y, the value of the lateral pressure from the concrete core σ_{cr}, and ratio of the column reinforcement.

2.4 Central compression strength

The literature review shows that obtaining a reliable formula for determining the strength of volumetric compressed concrete of CFST elements is not an easy task. Most often, empirical formulas, which have significant limitations depending on the conditions of carried out experiment, are used. In case of structural changes or the use of new types of concrete and steel grades, other formulas will be needed. In this case, it is necessary to correctly determine the lateral pressure of a steel tube σ_{cr} on concrete, which directly affects both the strength of the concrete f_{cc} and the stress in the tube σ_{pz}.

In this regard, it is important to obtain theoretically based, universal formulas for determining f_{cc}, σ_{pz} and the strength of CFST. The solution to this problem is proposed on the basis of the known strength function of volumetric compressed concrete [17]. In the case of uniform lateral pressure, the result of solving this function is Eq. (5) with a variable value k depending on the level of lateral pressure $m = \sigma_{cr}/f_{cc}$ and the type of concrete. For its determination, a formula is recommended

$$k = \frac{1 + a - am}{b + (c - b)m}, \tag{14}$$

where a, b are material coefficients determined based on experiments;

c is a parameter determining the nature of strength surface in the area of all-around compression (for a dense concrete core, the strength surface is open, and $c = 1$).

The average values of strength of normal concrete, calculated with a reliability of 50%, correspond to the coefficients $b = 0.096$ and $a = 0.5b$.

The analysis of relationship (14) shows that with high levels of sidework (with $m \to 1$), the value of the lateral pressure coefficient is $k \to 1$. In such cases, concrete destruction will be of shear nature, according to Coulomb's law. With the above-mentioned coefficients k for CFST, volumetrically loaded concrete destruction occurs due to combinations of break and shear, which corresponds to numerous experimental data.

Inserting the Eq. (14) into the Eq. (5) and performing some transformations, we will obtain:

$$f_{cc} = \alpha_c \gamma_c f_c; \tag{15}$$

$$\alpha_c = 0,5 + 0,75\overline{\sigma} + 0,25\sqrt{(\overline{\sigma} - 2)^2 + 16\overline{\sigma}/b}, \tag{16}$$

where $\overline{\sigma}$ is a relative value of the lateral pressure from the steel shell on the concrete core in limit state $\overline{\sigma} = \sigma_{cr}/(\gamma_c f_c)$.

Using the relationship (12) and performing some little manipulations, we can write the Eq. (12) as follows

$$\sigma_{pz} = \gamma_c f_c \left(\sqrt{\xi^2 - 3\overline{\sigma}^2} - \overline{\sigma} \right) \frac{A}{A_p}. \tag{17}$$

The formula for $\overline{\sigma}$ calculation is received from solving the task of determination of the maximum compression force received by a short centrally loaded column. Inserting (15), (16) and (17) into the Eq. (3), we obtain the following equation

$$N = \gamma_c f_c A \left[1 + \left(\frac{\overline{\sigma} - 2}{4} + \sqrt{\left(\frac{\overline{\sigma} - 2}{4} \right)^2 + \frac{\overline{\sigma}}{b}} - \frac{\overline{\sigma}}{2} + \sqrt{\xi^2 - 3\overline{\sigma}^2} \right) \right]. \tag{18}$$

It is obvious that the total axial force received by concrete and steel with standard cross-section depends only on relative lateral pressure $\overline{\sigma}$ with fixed values of geometry and design parameters of CFST (f_c, f_y, A, A_p). For illustrative purposes, **Figure 1** represents diagrams of changes of relative forces received by concrete $\overline{N_c}$ and the steel shell $\overline{N_p}$ and their sum $\overline{N}$ depending on $\overline{\sigma}$ value. All forces are determined here in relation to the destructive load.

Figure 1 shows that the graph of the total force change has a maximum point. The maximum compressive force can be found from the equation $\frac{d}{d\overline{\sigma}}(N(\overline{\sigma})) = 0$. After determining the derivative we have the equation

$$\left(\frac{b(\overline{\sigma} - 2) + 8}{\sqrt{b}\sqrt{b(\overline{\sigma} - 2)^2 + 16\overline{\sigma}}} - \frac{12\overline{\sigma}}{\sqrt{\xi^2 - 3\overline{\sigma}^2}} - 1 \right) = 0. \tag{19}$$

As a result of solving Eq. (19), the following formula was obtained

$$\overline{\sigma} = 0,48e^{-(a+b)}\xi^{0.8}. \tag{20}$$

Thus, the necessary formulas to calculate the strength of a short centrally loaded CFST have been received.

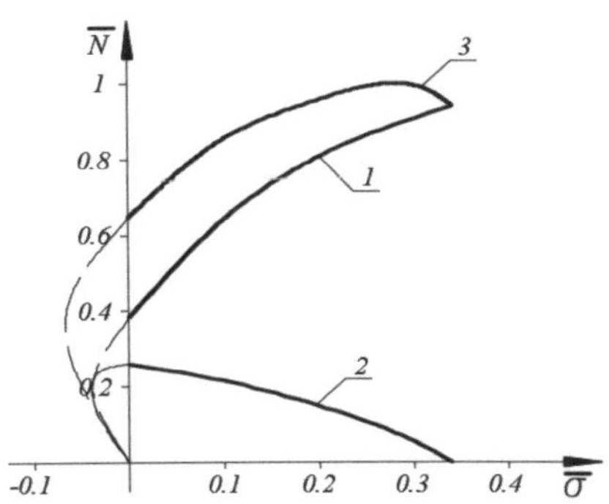

Figure 1.
Diagrams of changes of relative compressive forces received by concrete (1) and the steel shell (2) and their sum (3) depending on $\overline{\sigma}$ value.

2.5 Strength calculation of elements with spiral reinforcement

The construction of CFST columns can be improved by placing spiral reinforcement in the concrete core (**Figure 2**). This will have a positive effect on the strength and survivability of columns. A spiral, installed at some distance from the inner surface of the steel tube, can also increase the fire resistance of columns. Experimental studies [10, 11, 20] confirm the high efficiency of such structures.

The widespread practical use of reinforced CFST columns is constrained by the lack of reliable methods for determining their strength. In work [12], a numerical finite element analysis of the load resistance of compressed CFST elements with spiral reinforcement was carried out. But empirical formulas were used here to determine the strength of concrete and lateral pressure on concrete in the limiting state.

The strength of short centrally compressed reinforced CFST column can be determined by formula:

$$N_{u0} = f_{cc}A_c + \sigma_{pz}A_p + \sigma_s A_s, \tag{21}$$

where σ_s is the compressive stress in longitudinal reinforcement in the limiting state of element;

A_s is cross-sectional area of the longitudinal reinforcement.

Under the action of axial compressive force N, lateral pressure on the concrete takes place due to the restraining effect of the outer steel tube and spiral reinforcement. It is impossible to determine this pressure by the superposition principle, since the current problem is physically nonlinear. Therefore, the following calculation method is proposed.

First, the load resistance of a spirally reinforced concrete element that does not have an external steel tube is considered. As a result, the strength of concrete with confinement reinforcement f_{cs} is calculated. At the second stage of the calculation, the interaction of this element and the outer steel shell is taken into account.

To determine the strength of the concrete core f_{cs}, Eq. (15) and (16) are used with the replacement of $\overline{\sigma}$ by $\overline{\sigma}_{sc}$.

The value of relative lateral pressure $\overline{\sigma}_{sc}$ is calculated by the formula:

$$\overline{\sigma}_{sc} = \rho_{sc}\frac{\sigma_{sc}}{\gamma_c f_c}, \tag{22}$$

where ρ_{sc} is coefficient of confinement reinforcement by spirals;

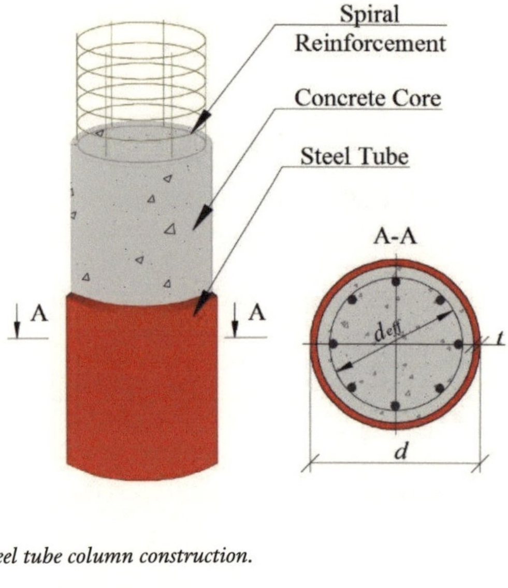

Figure 2.
Reinforce concrete filled steel tube column construction.

σ_{sc} is tensile stress in the spiral reinforcement, which can be determined from the formula:

$$\sigma_{s,c} = \varepsilon_{sc} E_{s,c} \leq f_{y,c}, \tag{23}$$

where ε_{sc} is tensile strain of spiral reinforcement;

$E_{s,c}$ is modulus of elasticity of steel of spiral reinforcement;

$f_{y,c}$ is yield point of steel of spiral reinforcement.

The following formula for calculating the value ε_{sc} by consecutive approximations is derived in the work [11]:

$$\varepsilon_{sc} = -\frac{\upsilon_{zr}}{q \nu_{cs} E_c} f_{cs}, \tag{24}$$

in which,

$$q = 1 - \frac{E_{s,c}}{E_c} \rho_{sc} (1 - \upsilon_{rr}), \tag{25}$$

ν_{cs} is the coefficient of elasticity at the maximum stress of concrete with confinement reinforcement.

The value ν_{cs} is calculated using the formula:

$$\nu_{cs} = \frac{f_{cs}}{\varepsilon_{cs} E_c}, \tag{26}$$

where ε_{cs} is the strain of concrete with confinement reinforcement at the maximum stress.

The values of coefficients of transverse deformations υ_{zr} and υ_{rr} are calculated using the formulas obtained in work [16]. The strain ε_{cs} is calculated using the formula obtained below.

Then the strength of spirally reinforced concrete core f_{cc1}, which has an outer steel shell, is determined. For this purpose the Eq. (15) is used, in the right-hand side of which the value $\gamma_c f_c$ is substituted by f_{cs}. The relative lateral pressure $\overline{\sigma}_1$ depends on constructional coefficient ξ_1, calculated by the formula:

$$\xi_1 = \frac{f_y A_p}{f_{cs} A}. \tag{27}$$

The lateral pressure on the concrete from the steel tube acts outside the diameter of the spiral d_{eff}. This pressure is calculated using the formula (20), but with the replacement of the coefficient ξ by ξ_2. Constructive coefficient ξ_2 is determined from the formula (10) when the strength of concrete is $\gamma_c f_c$.

Depending on $\overline{\sigma}_2$, the strength of the concrete of the peripheral zone f_{cc2} is calculated.

In order to simplify the calculations it is offered to use the averaged design compressive strength of concrete core f_{cc} for the method of limiting forces. It is determined from the formula:

$$f_{cc} = f_{cc2} (1 - \beta_c^2) + f_{cc1} \beta_c^2, \tag{28}$$

where β_c is the coefficient determined using the formula $\beta_c = d_{eff} / d_c$.

The stress σ_{pz} in the steel tube is calculated by the following formula:

$$\sigma_{pz} = \gamma_c f_c \left[\left(\xi_2^2 - 3\overline{\sigma}_m^2 \right)^{1/2} - \overline{\sigma}_m^2 \right] \frac{A}{A_p}, \tag{29}$$

in which $\overline{\sigma}_m$ – averaged value of relative lateral pressure of concrete core, calculated by the formula:

$$\overline{\sigma}_m = \overline{\sigma}_1 \beta_c + \overline{\sigma}_2 (1 - \beta_c). \tag{30}$$

The compressive stress in the longitudinal reinforcement σ_s should be determined from the condition of its combined deformation with the concrete core $\varepsilon_s = \varepsilon_{cz}$.

3. Deformation calculation of strength

3.1 General provisions

In a number of earlier published works it is shown that the most reliable calculations of the bearing capacity of CFST columns, taking into account their design features, can be carried out on the basis of nonlinear deformation model. The calculation sequence of similar designs for deformation model is in detail stated in [16].

The calculations are based on the assumptions specified in the EN 1992-1-1 standard. They are listed in the introduction. While processing the experimental data the values of random eccentricity are taken three times less than the values recommended by standards for design purposes. Thus, the centering of the samples along the physical axis is taken into account.

The calculation is based on the relationships between stresses and strains for the concrete core $\sigma_{cz} - \varepsilon_{cz}$, steel tube $\sigma_{pz} - \varepsilon_{pz}$ and reinforcement (if any) $\sigma_s - \varepsilon_s$. The concrete core and steel tube operate under conditions of volumetric stress state, which can change quantitatively and qualitatively with increasing load (**Figure 3**). The accuracy of calculations largely depends on the reliability of the adopted diagrams. At that, the diagrams contained in the regulatory documents are not suitable to evaluate the strength resistance of a concrete core and a steel shell. Therefore, reliable diagrams $\sigma_{cz} - \varepsilon_{cz}$ and $\sigma_{pz} - \varepsilon_{pz}$ need to be constructed initially. The form of a diagram set is recommended to be multipoint one. It is shown in [17] that such a method is the most universal one. At the second stage the strength of the compressed CFST is calculated.

3.2 The first stage of calculation

At the first stage, the deformation diagrams of the concrete core and the steel tube are constructed for the axial direction of the element. For this purpose, the load resistance of a short centrally compressed CFST element is considered. Load is imposed quickly. The concrete core is considered as a transversely isotropic body. The steel tube is considered to be an isotropic body. In the tube the stresses arise in the axial, circumferential and radial directions – σ_{pz}, $\sigma_{p\tau}$, σ_{pr}. The stress signs of the concrete core and steel shell depend on the load level. At low load levels, the value of the coefficient of transverse strains of steel exceeds the value of the coefficient of transverse strains of concrete. For these levels, there is no triaxial compression of concrete (**Figure 3b**). When the value of the coefficient of transverse strains of concrete exceeds the value of the coefficient of transverse strains of steel, the volumetric compression of concrete takes place (**Figure 3c**).

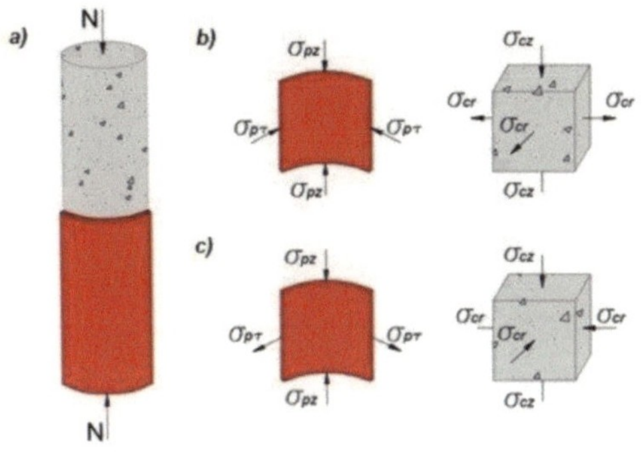

Figure 3.
Tension of steel tube and concrete core of the central compressed CFST column: a – scheme of loading; b – at low loading levels; c – at high loading levels.

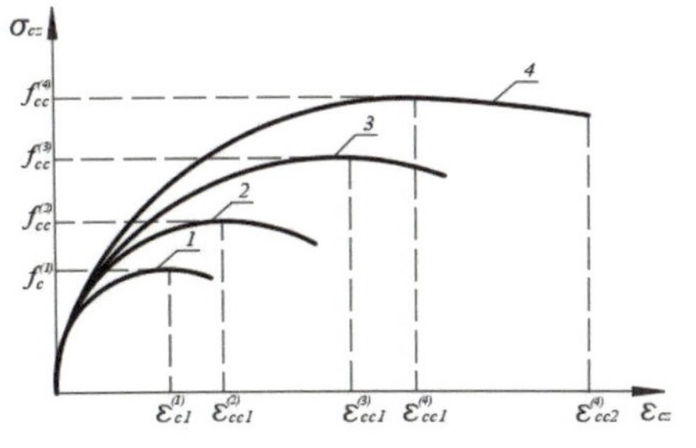

Figure 4.
Branch of concrete deformation charts at step-by-step strengthening of axial deformations: 1 - uniaxial compression, 2,3 - volume compression at the intermediate stages of deformation; 4 - volume compression in a limit state.

Curvilinear deformation diagrams are accepted for the concrete core. The coordinates of vertex of each diagram depend on the lateral pressure on the concrete from the steel tube. It is assumed that with an increase of the compressive force N, the lateral pressure on the concrete σ_{cr} goes up from zero to a certain limiting value. Therefore, the calculation requires the use of many such diagrams (**Figure 4**).

The coordinates of vertex of each diagram determine the strength of the concrete core (uniaxially compressed f_c or volumetrically compressed f_{cc}) and its strain (ε_{c1} or ε_{cc1} respectively).

There are many proposals for determining the strain ε_{cc1} in the literature. The main disadvantage of the above formulas is that they are all obtained from the results of the corresponding experiments. It greatly limits the scope of their application.

Let's show how one can get the corresponding formula based on the phenomenological approach.

Figure 5 shows the stress–strain diagram of compressed concrete, corresponding to the maximum reached stress and compare it with the uniaxial compressed concrete diagram. It follows from the above that the initial modulus of elasticity E_c for both diagrams is the same.

The strain ε_{cc1} at the vertex of the diagram $\sigma_{cz} - \varepsilon_{cz}$ is made up of elastic ε_{el} and plastic ε_{pl} components

$$\varepsilon_{\text{cc1}} = \varepsilon_{\text{el}} + \varepsilon_{\text{pl}}. \tag{31}$$

Elastic strain ε_{el} is associated with the elastic part of the strain of uniaxially compressed concrete ε'_{el} by the following relationship:

$$\varepsilon_{\text{el}} = \varepsilon'_{\text{el}} \frac{f_{\text{cc}}}{\gamma_c f_{\text{c}}}. \tag{32}$$

Plastic strain ε_{pl} is associated with the plastic part of the strain of uniaxially compressed concrete ε'_{pl} by a similar relationship:

$$\varepsilon_{\text{pl}} = \varepsilon'_{\text{pl}} \left(\frac{f_{\text{cc}}}{\gamma_c f_{\text{c}}} \right)^m, \tag{33}$$

where m is the exponent, $m > 1$.

The parameter m takes into account the fact that the increase in strains of volumetrically compressed concrete is more intense than the increase in its strength.

Thus, the total deformation of the volume-compressed concrete at the maximum stress is determined by the formula

$$\varepsilon_{cc1} = \varepsilon_{c1} \alpha_c^m \left[1 - \frac{\gamma_c f_c}{\varepsilon_{c1} E_c} \left(1 - \alpha_c^{(1-m)} \right) \right]. \tag{34}$$

The performed statistical analysis showed that the best match with the results of the experiments corresponds to a value of m, calculated by the formula

$$m = 1.7 + \frac{3.5}{\sqrt{\gamma_c f_c}}, \tag{35}$$

where f_c is in MPa.

According to the recommendations of [21] the ultimate strain of a volume-compressed concrete is determined by the formula

$$\varepsilon_{\text{cc2}} = \varepsilon_{\text{c2}} \frac{\varepsilon_{\text{cc1}}}{\varepsilon_{\text{c1}}}, \tag{36}$$

where ε_{c2} is the ultimate strain for uniaxial compressed concrete.

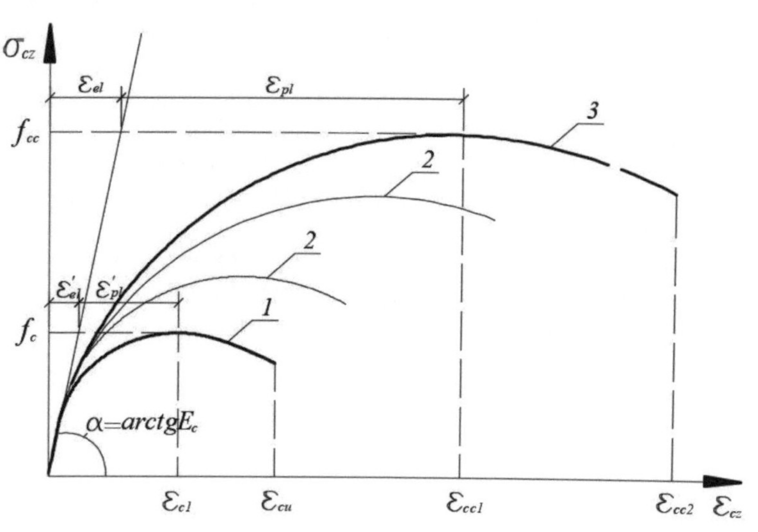

Figure 5.
The graphs of deformation for uniaxial compressed (1) and volume-compressed (2,3) concrete.

When coordinates of parametric points of the deformation charts of volumetrically compressed concrete are known, it is possible to calculate the bearing capacity of CFST columns based on the deformation model analysis.

To construct the diagrams $\sigma_{cz} - \varepsilon_{cz}$ and $\sigma_{pz} - \varepsilon_{pz}$, a step-by-step increase in the strains of the concrete core and steel tube is carried out while ensuring the condition $\varepsilon_{cz} = \varepsilon_{pz}$. All components of the stress–strain state of concrete and steel are calculated at each j-th step.

The analytical relationship between strains and stresses for any point of the concrete core is written in the form of a system of equations:

$$\begin{Bmatrix} \varepsilon_{cz} \\ \varepsilon_{cr} \end{Bmatrix} = \frac{1}{E_c} \times \begin{bmatrix} \nu_{cz}^{-1} & -2\upsilon_{zr}\nu_{ci}^{-1} \\ -\upsilon_{zr}\nu_{ci}^{-1} & \left(\nu_{cr}^{-1} - \upsilon_{rr}\nu_{ci}^{-1}\right) \end{bmatrix} \times \begin{Bmatrix} \sigma_{cz} \\ \sigma_{cr} \end{Bmatrix}. \tag{37}$$

The elastic–plastic properties of concrete are taken into account by the coefficients of elasticity ν_{ck} $(k = z, r, i)$ and variable coefficients of transverse strains υ_{zr}, υ_{rr}. The subscripts z and r are used for axial and transverse directions, and the subscript i is used for the coefficient of elasticity depending on the intensity of stress and intensity of strain.

The values of the intensity of stresses and strains are calculated using the well-known formulas of solid mechanics. Using the coefficients of elasticity ν_{ci} and transverse strains υ_{zr}, υ_{rr}, the strains along one direction (axial or transverse) depending on the stresses of the other direction are calculated in a matrix of system pliability (37).

The stress state of a steel tube obeys the hypothesis of a uniform curve [22]. In accordance with this hypothesis, the dependence σ_{pi}-ε_{pi}, obtained under uniaxial tension, is accepted for complex stress states. Here σ_{pi} is the intensity of stresses, and ε_{pi} is the intensity of strains.

The initial diagram "$\overline{\sigma}_p - \overline{\varepsilon_p}$" is recommended to be tri-linear (rules of Russia - CR 266.1325800.2016). However, when modelling steel sections under a complex stress state, it is advisable to use a deformation diagram calculated using the generalized parameters $\overline{\sigma}_{pi} = \sigma_{pi}/f_y$ and $\overline{\varepsilon}_{pi} = \varepsilon_{pi}E_p/f_y$ (**Figure 6**). The coordinate values of the characteristic points of the generalized diagram can be taken from **Table 1**.

Communication between strains and stresses for any point of an external steel shell in elastic and elasto-plastic stages can be presented the following equations system:

$$\begin{Bmatrix} \varepsilon_{pz} \\ \varepsilon_{p\tau} \\ \varepsilon_{pr} \end{Bmatrix} = \frac{1}{\nu_p E_p} \times \begin{bmatrix} 1 & -\upsilon_p & -\upsilon_p \\ -\upsilon_p & 1 & -\upsilon_p \\ -\upsilon_p & -\upsilon_p & 1 \end{bmatrix} \times \begin{Bmatrix} \sigma_{pz} \\ \sigma_{p\tau} \\ \sigma_{pr} \end{Bmatrix}. \tag{38}$$

Here σ_{pz}, $\sigma_{p\tau}$, σ_{pr} are normal (main) stresses in a tube in the axial, circumferential and radial directions; ε_{pz}, $\varepsilon_{p\tau}$, ε_{pr} are strains of a steel tube in the corresponding directions; E_p is the initial module of tubes elasticity; ν_p is coefficient of steel elasticity; υ_p is coefficient of tubes cross strain.

The stresses and strains acting on the principal planes are used in Eqs. (37) and (38). Experiments show [16] that in the stage of yield Chernov-Luders lines appear on the surface of the steel tube. These lines are angled 45° to the longitudinal axis of the CFST. Therefore, shear stresses and shear strains are equal to zero here.

The stress–strain states of the concrete core and steel tube largely depend on the values of the coefficients of transverse strain and the coefficients of elasticity of the materials. Therefore, their reliable determination is very important when

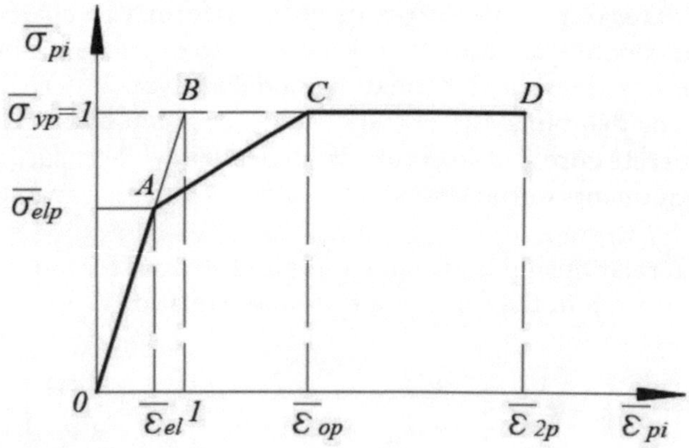

Figure 6.
Generalized calculation diagram of steel, operating under conditions of complex stress state.

Parameter of diagram	Steel classes according to the set of rules Russia - SP 16.13330.2018					
	S245, S255	S285	S345, S345K, S375	S390	S440	S590, S590K
$\overline{\varepsilon}_{el}$	0,80	0,80	0,80	0,90	0,90	0,90
$\overline{\sigma}_{elp}$	0,92	0,92	0,92	1,00	1,00	1,00
$\overline{\varepsilon}_{op}$	1,70	1,70	1,70	1,70	1,70	1,70
$\overline{\sigma}_{yp}$	1,00	1,00	1,00	1,00	1,00	1,00
$\overline{\varepsilon}_{2p}$	14,0	15,0	16,0	17,0	17,0	18,0

Table 1.
Coordinates of characteristic points of the generalized steel deformation diagram, constructed in the axes $\overline{\sigma}_{pi} - \overline{\varepsilon}_{pi}$.

calculating the strength of CFST columns. Formulas for calculating these coefficients are given in work [16].

The solution of the Eqs. (37) and (38), taking into account the joint deformation of concrete and steel tube, allows obtaining the formula for calculating the lateral pressure

$$\sigma_{cr} = \frac{\left(\upsilon_p - \upsilon_{zr}\frac{d_c}{d_c+\delta}\frac{\nu_{cz}}{\nu_{ci}}\right)\varepsilon_{cz}}{K_p + K_c}, \tag{39}$$

in which K_p and K_c are the parameters defining condition of steel shell and concrete core.

$$K_p = \frac{0,5\upsilon_p}{\nu_p E_p}\left[\upsilon_p\left(\frac{d}{\delta} - 1\right) - \left(\frac{d}{\delta} + 1\right)\right]; \tag{40}$$

$$K_c = \frac{\beta_r}{\nu_{ci}E_c}\left(\frac{2\upsilon_{zr}^2\nu_{cz}}{\nu_{ci}} + \upsilon_{rr} - \frac{\nu_{ci}}{\nu_{cr}}\right). \tag{41}$$

When the strain ε_{cz} and lateral pressure σ_{cr} are known, all other components of the stress–strain state of CFST column can be calculated. The strains are incrementally increased until the stress σ_{cz} reaches the strength of volumetrically compressed concrete $f_{cc}^{(n)}$ (**Figure 4**), previously calculated using formulas (15) and (16). The calculation is performed on a computer.

After that we compare the last value of strain ε_{cz} with the strain calculated on a formula (34) $\varepsilon_{cc1}^{(n)}$ in top of the deformation chart of concrete. In the existing incoherence $\left|\varepsilon_{cz} - \varepsilon_{cc1}^{(n)}\right| > \Delta_\varepsilon$ (Δ_ε– the accuracy of calculations set by the estimator) we specify value of an exponent m in a Eq. (34) and repeat all calculations.

Upon termination of calculations we receive arrays of numerical data for deformation charting of concrete core $\{\varepsilon_{cz}\} - \{\sigma_{cz}\}$ and steel shell $\{\varepsilon_{pz}\} - \{\sigma_{pz}\}$.

3.3 The second stage of calculation

At the second stage, the bearing capacity of the eccentrically loaded CFST element is calculated. The design scheme of the normal section of element is shown in **Figure** 7.

In the calculation process, the deformation of the most compressed fiber of the concrete core $\varepsilon_{c\,\max}$ is increased step-by-step. At each step, using the Bernoulli hypothesis, the diagram of strain fiber of the steel tube $\varepsilon_{p\,\min}$. The search for this value is carried out with a gradual shortening strain decrease (starting from $\varepsilon_{c\,\max}$) or the build-up of the elongation strain $\varepsilon_{p\,\min}$ (starting from zero).

The normal section of the calculated element is conditionally divided into small sections with areas of concrete A_{ci} and steel shell A_{pk}. In the presence of longitudinal reinforcement, the cross-sectional area of each bar is designated as A_{sj}.

The origin of coordinates is aligned with the geometric center of the element's cross section. If the Bernoulli hypothesis is observed, there is a strain in the center of each section of concrete and steel tube. With known strains, the corresponding stresses are determined according to the results of the first stage of the calculation. The stresses are assumed to be evenly distributed within each section of concrete

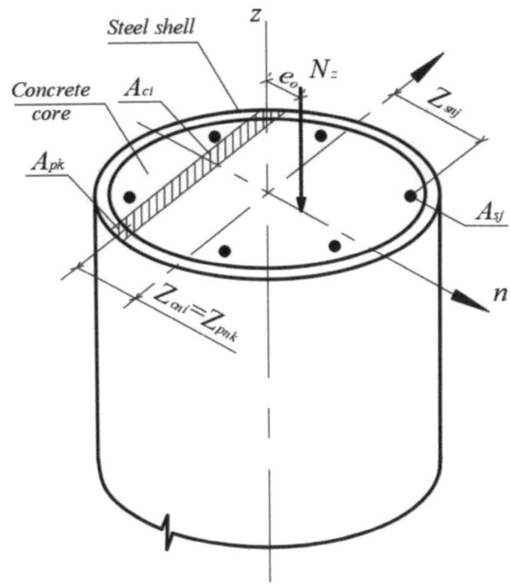

Figure 7.
Design model of the normal section of the CFST element deformations of the normal cross section is designed, corresponding to the equilibrium condition of the calculated element. In order to develop such a diagram it is required to find the corresponding value of the strain of the least compressed (stretched).

and steel tube. After each step of strain $\varepsilon_{c\max}$ increasing, it is necessary to ensure that the equilibrium conditions are met:

$$N_z = \sum_i \sigma_{czi} A_{ci} + \sum_k \sigma_{pzk} A_{pk} + \sum_j \sigma_{sj} A_{sj}; \tag{42}$$

$$N_z e_0 = \sum_i \sigma_{czi} A_{ci} Z_{cni} + \sum_k \sigma_{pzk} A_{pk} Z_{pnk} + \sum \sigma_{sj} A_{sj} Z_{snj}, \tag{43}$$

in which Z_{cni}, σ_{czi} – the coordinate of the gravity center of the i-th section of concrete and the stress of the axial direction at the level of its gravity center; Z_{pnk}, σ_{pzk} – the coordinate of the gravity center of the steel shell k-th section and the stress of the axial direction at the level of its gravity center; Z_{snj}, σ_{sj} – the coordinate of the gravity center of the of the longitudinal reinforcement j-th bar and the stress in it.

When both equilibrium conditions are met, the value of the compressive force N_z corresponding to the given strain $\varepsilon_{c\max}$ is fixed. Next, the strain of the most compressed fiber of the concrete core increases and all calculations are repeated. The limiting values of this strain ε_{ccu} can be accepted according to the recommendations [21].

The problem of determining the strength reduces to finding the value of the strain of the most compressed fiber $\varepsilon_{c\max} \leq \varepsilon_{cc1}$, corresponding to the maximum value of the compressive longitudinal force N_u. The calculation results show that under certain design parameters of CFST columns, the strain $\varepsilon_{c\max}$ does not reach ε_{cc1}. Then the stress in the concrete σ_{cz} cannot achieve its strength at triaxial compression. This design situation occurs when using low strength concrete and a strong steel shell with a small ratio d/δ. Therefore, the criterion for the loss of the strength of the column is the achievement of the maximum value of compressive force in the process of increasing the strain of the most compressed fiber.

The proposed method makes it possible to limit the axial strains of the columns. It is known from experiments that the strain of compressed CFST elements can reach 5 ÷ 10% [16]. With such strains, the operation of the columns of the buildings becomes impossible. Thus, excessive strain can determine the ultimate limit state of the CFST column. The maximum permissible values of these strains can be set by a structural engineer, depending on a specific design situation for a designed building or a structure.

3.4 Calculation of flexible elements

Due to the complex nature of load resistance of CFST columns, in design practice, as a rule, the simplified methods of calculation of their bearing capacity are used. At that, flexibility is usually taken into account by the coefficient of longitudinal bending, determined according to empirical relationships. In the monograph we consider the deformation calculation of CFST column bearing capacity.

A rod of a circular cross-section with a constant length, loaded by a compressive force N applied to the ends with the same initial eccentricity e_0 (no less accidental than e_a) and hinged at its ends is regarded as a basic case. The deformation scheme of such a rod is shown on **Figure 8**.

According to the known positions of structural mechanics, if we apply force N along the axis that coincides with the physical gravity center of an elastic rod cross-section, the rod will remain a rectilinear one until the force reaches the value of the critical load N_u corresponding to the moment of stability loss. Only after that the middle part of the rod will receive the corresponding deflection f_{cr}.

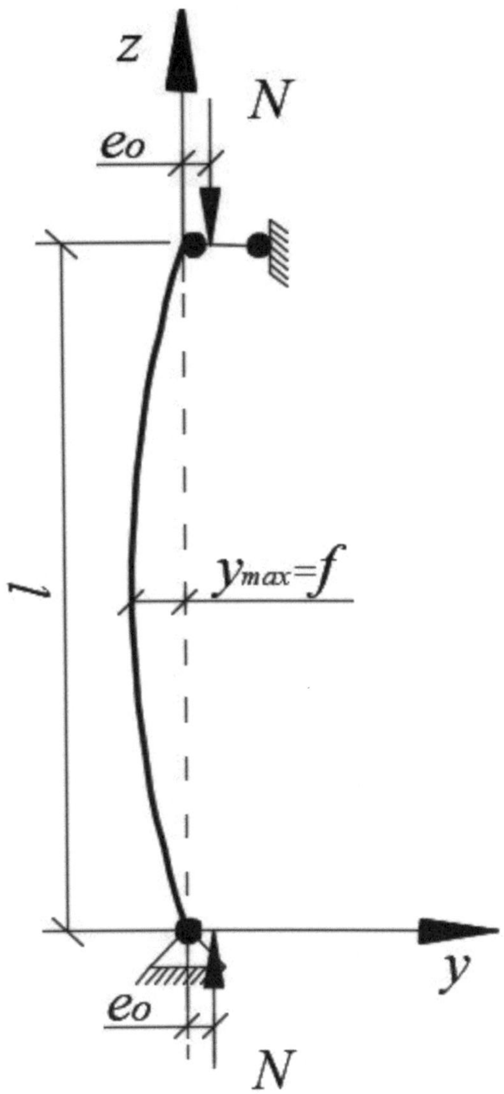

Figure 8.
The scheme of a compressed rod deformation.

A bending moment M will appear in any section along the length of the bar from the compressive force N. The moment M is calculated by the formula

$$M = N(e_0 + y), \tag{44}$$

where y is the horizontal displacement value of the cross-section in question.

With the increase of the bending moment, the strength of a compressed rod normal section decreases, which must be taken into account during the calculation. On the other hand, the axial load increase to a critical value in the columns of great flexibility can lead to a very significant increase of transverse deformations - the loss of stability of the second kind. With a certain transverse deflection, the compressive load reaches a maximum value, after which its decrease is observed with a further deflection increase (**Figure 9**). At the same time, the strength properties of materials from which the column is made will not be implemented fully.

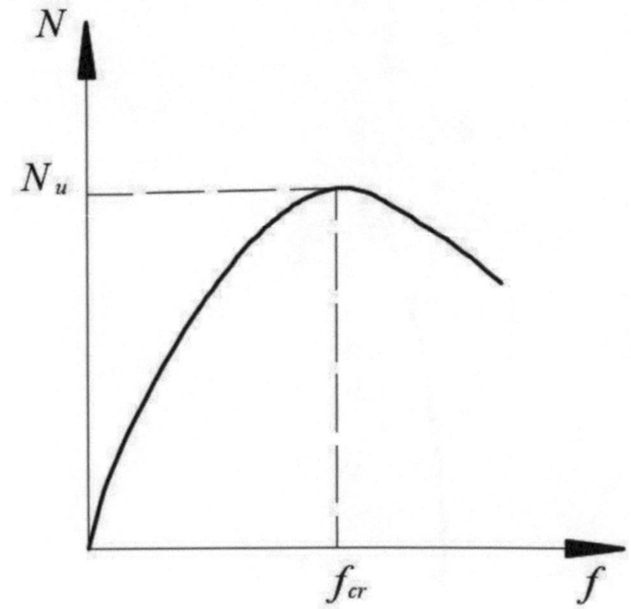

Figure 9.
The dependence of compressive force on deflection

The main assumptions that are directly relevant to this study are the following ones:

- the calculation is based on the theory of small displacements;
- the shear deformations are neglected in comparison with the bending deformations of the rod axis;
- the distribution of deformations along a cross section corresponds to the hypothesis of plane cross sections.

The flexibility of the column is determined for the reduced cross-section. For the base case under consideration, this flexibility can be approximated by the following formula:in which l is the estimated length of the rod; $(EI)_{eff}$, $(EA)_{eff}$ are effective stiffness of the most loaded reduced section for bending and compression.

$$\lambda_{eff} = l \cdot \sqrt{\frac{(EA)_{eff}}{(EI)_{eff}}}. \tag{45}$$

It is recommended to calculate the stiffness $(EI)_{eff}$ and $(EA)_{eff}$ in the first approximation by the following formula:

$$(EI)_{eff} = 0,5E_cI_c + 0,5E_pI_p + E_sI_s; \tag{46}$$

$$(EA)_{eff} = 0,5E_cA_c + 0,5E_pA_p + E_sA_s, \tag{47}$$

where I_c, I_p, I_s are the moments of inertia of a concrete core, a steel tube and a longitudinal reinforcement; E_c, E_p, E_s are the moduli of elasticity of concrete, steel case and longitudinal reinforcement.

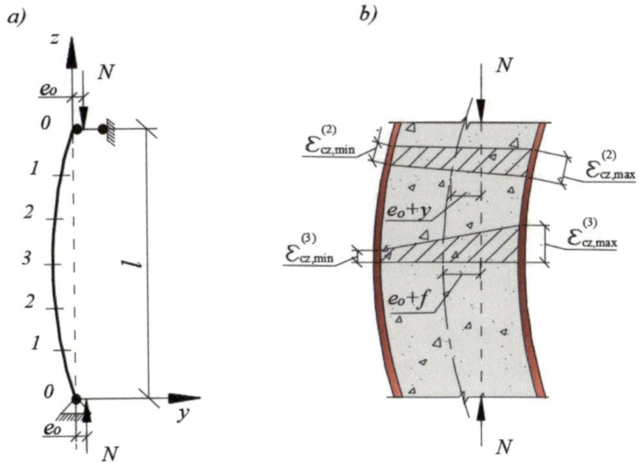

Figure 10.
The design scheme of a flexible pipe-concrete column: a - the decomposition of the compressed rod along the length; b - distribution diagrams of concrete relative deformations in Section 2 and 3.

Flexibility can have a significant effect on the load capacity of compressed elements when the condition $\lambda_{eff} > \lambda_0$ is performed, in which the threshold value of flexibility is calculated by the following formula

$$\lambda_0 = \pi\sqrt{0,01(EA)_{eff}/N_{u0}}, \tag{48}$$

where N_{u0} is the strength of a short, centrally compressed CFST element.

The compressive stress in the longitudinal reinforcement σ_s is determined from the condition of its joint deformation with the concrete core. This takes into account the limitation $\sigma_s \leq f_{y,s}$.

The calculation is based on the step-iteration method. During the second stage, an eccentrically loaded compressed element is divided along its length into n equal segments, at that $n \geq 6$ (**Figure 10**). Normal sections at the end of each segment are divided into small sections conventionally with the areas of concrete A_{ci} and steel A_{pk} tube.

The area of one rod of longitudinal reinforcement is A_{sj}. Then the calculation process is performed in the following sequence. First, only one normal cross-section of a rod is considered, in which the maximum bending moment arises. This cross section is located in the middle of the column height for the articulated column loaded by a compressive force N with the initial eccentricity e_0. The strain of the most compressed fiber of the concrete core $\varepsilon_{cz\max}$ is increased stepwise in this section.

At each step, the relative deformation of the least compressed (stretched) fiber $\varepsilon_{cz\min}$ is determined, corresponding to the conditions of equilibrium cross section. The equilibrium conditions are written in the form of the following equation system:

$$N = (EA)_{eff}\varepsilon_0; \tag{49}$$

$$N(e_0 + f) = (EI)_{eff}\frac{1}{r}, \tag{50}$$

where N is the longitudinal compressive force corresponding to the accepted deformation diagram; ε_0 is a fiber relative deformation located at the gravity center of calculated section; f is the deflection at the point of maximum bending moment; $\frac{1}{r}$ is the curvature of the longitudinal axis in the considered cross-section, determined by the following formula

$$\frac{1}{r} = \frac{\varepsilon_{cz\,\max} - \varepsilon_{cz\,\min}}{d - 2\delta}. \tag{51}$$

Cross-section stiffnesses $(EA)_{eff}$ and $(EI)_{eff}$ are found taking into account the corresponding elastic coefficients of concrete and steel [7].

The effect of longitudinal bending is taken into account via the eccentricity of the longitudinal force increase by the amount of rod deflection f in the calculated section. In the first approximation, the deflection value is determined depending on the curvature of the calculated normal section. Taking into account the dependence (51), we can write the following formula

$$f = \frac{l^2}{\pi^2} \frac{\varepsilon_{cz\,\max} - \varepsilon_{cz\,\min}}{d - 2\delta}, \tag{52}$$

where l is the estimated length of the considered rod.

An improved deflection value f should be found at each calculation step for a more reliable calculation of a compressed rod longitudinal bending. This can only be done by adjusting the stiffness along a rod length.

The numerical solution of the problem of calculating the deflection [16] with the number of partitions n = 6 allows us to obtain the following formula

$$f = \frac{l_0^2}{266}\left(\frac{1}{r_0} + 6\frac{1}{r_1} + 12\frac{1}{r_2} + 8\frac{1}{r_{\max}}\right), \tag{53}$$

where $\left(\frac{1}{r_0}\right)$ is the curvature of the element on the upper (lower) supports; $\left(\frac{1}{r_i}\right)$ is the curvature of the element in the i-th section; $\left(\frac{1}{r_{\max}}\right)$ is the curvature in the middle of the height.

The problem under consideration is solved as follows. The deviations y of the longitudinal axis of the compressed rod from the vertical are calculated in the sections at the boundaries of each segment into which an element is divided with the deflection found in the first approximation according to the formula

$$y = f \sin(\pi z / l). \tag{54}$$

Then the distribution of the relative deformations is established for these cross-sections, using the Eqs. (49) and (50) and by the replacement of f into y. Moreover, during the determination of $\varepsilon_{cz\,\max}$ and $\varepsilon_{cz\,\min}$ for each section, it is necessary to satisfy two conditions:

- the equilibrium of the normal section, i.e. the observance of equalities by the Eqs. (49) and (50);
- the constancy of the longitudinal force value, which is assumed to be the same as for the mean most stressed section.

Let's note that the stiffness characteristics $(EA)_{eff}$ and $(EI)_{eff}$ depend on the parameters of the strain diagram. Therefore, they will be different for each section.

After the determination of $\varepsilon_{cz\,\max}$ and $\varepsilon_{cz\,\min}$ according to the Eq. (51), the curvatures in the support and intermediate sections of the rod are found, and by the Eq. (53) the deflection f is specified. The process of deflection refinement can be repeated until a predetermined calculation accuracy is achieved.

They record the value of the compressive longitudinal force N for the assumed value of the relative strain of the most compressed fiber of the concrete core $\varepsilon_{cz\,\max}$ of the average cross-sectional rod and the refined deflection f. Then the strain of the most compressed fiber of the concrete core $\varepsilon_{cz\,\max}$ is increased and the whole procedure of calculations is repeated. Thus, the dependence "$N - f$" is developed (see **Figure 9**). The maximum value of the longitudinal force N_u, perceived by the rod, is taken as the bearing capacity.

4. Comparison of calculated bearing capacity with experimental data

According to the proposed method, the algorithm for estimate the stress–strain state and calculate the load-bearing capacity of compressed concrete filled steel tube elements was developed and this algorithm was implemented in the computer program. The results of the calculations are compared with the experiment data of CFST samples made of normal concrete. These data were obtained by many researchers for 569 experiments with short centrally compressed columns, 512 flexible centrally compressed columns and 292 eccentrically compressed elements.

Experimental data was taken from research works [16, 23, 24].

In order to obtain more objective information, the experimental data of samples were analyzed with a large range of geometric and structural parameter variation:

- an outer diameter of an outer steel shell – d = 89 ÷ 1020 mm;
- the thickness of an outer steel shell wall – δ = 0.8 ÷ 13.3 mm;
- the yield point of a shell steel – R_s = 165.8 ÷ 853 MPa;
- the prismatic strength of the initial concrete – R_b = 11.7 ÷ 127 MPa;
- various concretes (normal, ultrahigh-strength, pre-stressing);
- length to diameter ratio l/d =2÷49;
- the relative eccentricity of the longitudinal force e_0/d =0÷0.94.

The results of the comparison show a completely satisfactory coincidence of experimental destructive loads with theoretical values (**Table 2**).

The data in **Table 2** show a good agreement between theory and practice.

According to the results of the data of work [23], the calculations according to Eurocode 4 (EN 1994-1-1: 2004) have a slightly worse accuracy. However, the main advantage of the proposed calculation method is its versatility. In particular, when using this method, one can take into account the presence of a high-strength rod and (or) spiral reinforcement, the effect of preliminary lateral compression of the concrete core [16]. The research work [13] verified the acceptability of the EN 1994-1-1: 2004 method for calculating the strength of compressed CFST made of various types of concrete: normal, ultrahigh-strength, self-compacting, light-weight concretes and engineered cementitious composite. It is concluded that the

Type of tested elements	No of tests	Average Test/Calculate	Stand. Deviation Test/Calculate
Short No Moment	569	1.04	0.068
Long No Moment	512	1.08	0.077
Long and Short with Moment	292	1.06	0.072
The overall	1373	1.07	0.073

Table 2.
Summary of Comparison of Calculated Bearing Capacity with Experimental Data.

calculation accuracy is satisfactory only for normal concrete. The proposed method makes it possible, with an appropriate selection of the material coefficients a and b in Eqs. (14), (16) and (20), to provide the required accuracy of calculations.

Based on the results of the carried out analysis, the following values of the coefficients of materials for various types of concrete can be recommended:

- for fine grained and for ultrahigh-strength concrete – b=0.13 and a=0.5b;
- for self-compacting concrete – b = 0.098, a =0.5 b;
- for lightweight concrete and for engineered cementitious composite – b = 0.3, a = 0.5 b.

Given recommendations are preliminary and need to be clarified, since they have been obtained on the basis of processing a very limited amount of experiments.

5. Discussion

The analysis of the results of the carried out researches shows that there are very significant advantages of the nonlinear deformation model in comparison with the currently used methods for calculating the bearing capacity of CFST columns. The proposed calculation method takes into account the complex stress state of the concrete core and steel tube, which is constantly changing with increasing load, and the physical and geometric nonlinearity of the structure. In the course of the calculation, it is possible to obtain a clear picture of the stress–strain state of the structure at various stages of loading.

The main dependences for finding the strength and strain characteristics of a concrete core and a steel tube are obtained phenomenologically. They correspond to the basic principles of solids mechanics. The resulting formulas are more universal than empirical dependencies. For example, they are true for different types of concrete. In principle, the developed method is applicable for calculating the bearing capacity of composite columns with various cross-sectional shapes and various variants of reinforcement of a concrete core. Differences in designs are easily taken into account when developing calculation algorithms for specific tasks.

The use of a multi-point method for constructing the diagrams of concrete deformation allows improving the accuracy of calculations. Previously, these diagrams were accepted either for uniaxially compressed concrete, or for volumetrically compressed concrete at the stage of ultimate equilibrium of the structure. In the first case the value of the bearing capacity turned out to be underestimated, and in the second case - overestimated.

The proposed criterion for achieving the bearing capacity of CFST columns is important for practical calculations. The use of this criterion makes it possible to identify the cases when the strength properties of a concrete core cannot be fully used. Calculation by the method of limiting efforts does not always reflect the physical essence of the process and can lead to significant errors.

From the point of view of modern concepts of solid mechanics, steel-reinforced concrete structures refer to nonlinear and non-equilibrium deformable systems. The feature of such system calculation is the need to refine the values of the existing forces and displacements consistently, since the internal forces and the rigidity of the structures are interdependent.

The proposed method of CFST load capacity calculation allows to take into account these features. Considering flexibility the higher stiffness of the compressed rod is taken into account at the sites located closer to its supports. In this regard, it is obvious that the correct implementation of this method in practice will allow to obtain more reliable calculation results in comparison with the currently used semi-empirical approach.

Besides, this method makes it possible to perform the calculations of normal cross section and stability strength from a unified point of view. During the calculation, it is possible to track (in terms of longitudinal deformation value) the completeness of concrete and steel strength property use. If the material deformations reach the maximum permissible values, it can be concluded that the strength of the structures is lost. If this is not observed in the loss of the load-bearing capacity of the structure, a conclusion can be made about the loss of stability of the second kind.

It is especially important, that the proposed method with an appropriate refinement can be used for calculating the compressed structures made of various constructional materials.

One more important circumstance should be noted. It is known that in CFST columns, even before the onset of complete loss of bearing capacity, axial deformations can reach excessively large values at which the operation of real structures becomes impossible. In these cases, the limiting deformation can become dominant, determining ULS. In this regard, during the calculation of bearing capacity the axial deformations of the compressed CFST elements should be limited. This approach can be implemented only when calculating with the use of a nonlinear deformation model of reinforced concrete.

The proposed method can be effectively used to calculate long-term load columns [25].

6. Conclusions

A new technique to determine the strength of compressed CFST was proposed. Based on the known principles of deformation calculation, it takes into account the specific features of CFST adequately. The methodology uses new dependencies to determine the strength and the ultimate deformation of a concrete core, as well as the way of concrete deformation diagram development. It allows to perform the combined calculation of CFST strength, taking into account their flexibility and the calculation of possible stability loss. There is no need for an empirical formula to determine the critical force proposed by modern design standards for composite structural steel structures in the practical application of the method.

The versatility of this method should be emphasized separately. The method is acceptable for CFST columns made of various types of concrete using various technologies.

The practical use of the proposed method gives a reliable estimate of the stress–strain state and the strength of concrete filled steel tube columns.

Author details

Anatoly Krishan
Department of Building Design and Constructions, Nosov Magnitogorsk State Technical University, Magnitogorsk, Russia

*Address all correspondence to: kris_al@mail.ru

References

[1] Bergmann R, Matsui C, Meinsma C, Dutta D. Design guide for concrete field hollow section columns under static and seismic loading. Koln: Verlag TUV Rheinland; 1995. 68 p.

[2] Chen B. New development of long span CFST arch bridges in China. Long span CFST arch bridges. In: Chinese-croatian joint colloquium; Brijuni islands; 2008. p. 357-368.

[3] Han L-H, Li W, Bjorhovde R. Developments and advanced applications of concrete filled steel tubular (CFST) structures. Journal of Constructional Steel Research. 2014; 100:211-228. DOI: 10.1016/j.jcsr.2014.04.016

[4] Nishiyama I, Morino S, Sakino K, Nakahara H. Summary of Research on Concrete-Filled Structural Steel Tube Column System Carried Out Under the US, JAPAN Cooperative Research Program on Composite and Hybrid Structures. Building Research Institute Japan; 2002. 176 p.

[5] Shanmugam N E, Lakshmi B. State of the art report on steel-concrete composite columns. Journal of Constructional Steel Research. 2001; 57 (10):1041–1080. DOI: 10.1016/S0143-974X(01)00021-9

[6] Bhure N, Tiwari N. Steel Concrete Composite Construction - A Review. IJRASET. 2018; 6:564–566. DOI: 10.22214/ijraset.2018.11089

[7] Krishan A L, Astafeva M A, Chernyshova E P. Strength Calculation of ShortConcrete Filled Steel Tube Columns. Int. Journal of Concrete Structures and Materials. 2018; 12:84. DOI:10.1186/s40069-018-0322-z

[8] Pattel V I, Hassanein M F, Thai H T, Al Abadi H, Elchalakani M, Yai B. Ultra-high strength circular short CFST columns: Axisymmetric analysis, behaviour and design. Journal Engineering Structures. 2019;179: 268-283. DOI:10.1016/j.engstruct.2018.10.081

[9] Xu L, Zhou P, Chi Y, Huang L, Ye J, Yu M. Performance of the High-Strength Self-Stressing and Self-Compacting Concrete-Filled Steel Tube Columns Subjected to the Uniaxial Compression. International Journal of Civil Engineering. 2018;16(9): 1069-1083. DOI: 10.1007/s40999-017-0257-9

[10] Hamidian M R, Jumaat M Z, Alengaram U J, Ramil Sulong N H, Shafig P. Pitch spasing effect on the axial compressive behavior of spirally reinforced concrete-filled steel tube (SRCFT).Thin-Walled Structures. 2016; 100: 213–223. DOI:10.1016/j.tws.2015.12.011

[11] Krishan A L, Troshkina E A, Astafeva M A. Strength of Short Concrete Filled Steel Tube Columns with Spiral Reinforcement. IOP Conf. Series: Materials Science and Engineering. 2017;262: 1-7. DOI: 10.1088/1757-899X/262/1/012048

[12] Xiamuxi A, Hasegawa A. A study on axial compressive behaviours of reinforced concrete filled tubular steel column. Journal of Constructional Steel Research. 2012;76:144–154. DOI: 10.1016/j.jcsr.2012.03.023

[13] Hossain K M A, Chu K. Confinement of six different concretes in CFST columns having different shapes and slenderness. International Journal of Advanced Structural Engineering. 2019;11:255-270. DOI: 10.1007/s40091-019-0228-2

[14] Krishan A L, Rimshin V I, Troshkina E A. Experimental Research of the Strength of Compressed Concrete Filled

Steel Tube Elements. Advances in Intelligent Systems and Computing. VIII International Siberian Transport Forum —TransSiberia. Springer. 2019; 2: 560-566. DOI: 10.1007/978-3-030-37919-3

[15] Hassanein M F, Patel V I, Elchalakani M, Thai T-H. Finite element analysis of large diameter high strength octagonal CFST short columns. Thin-Walled Structures. 2018;123:467-482. DOI: 10.1016/j.tws.2017.11.007

[16] Krishan A L, Rimshin V I, Astafeva M A. Szhatie trubobetonnie elementi. Teoriya I praktika. Moskva: ASV; 2020. 375 p. (rus).

[17] Karpenko N I. Obshchie modeli mekhaniki zhelezobetona. Moscow: Stroiizdat; 1996. 416 s. (rus).

[18] Saatcioglu M, Razvi S R. Strength and ductility of confined concrete. Journal of Structural Engineering. 1992; 118(6):1590-1607. DOI: 10.1061/(asce) 0733-9445(1992)118:6(1590)

[19] Mander J B, Priestley M J N, Park R. Theoretical stress-strain model for confined concrete. Journal of Structural Engineering. 1988;114(8):1804-1826. DOI: 10.1061/(asce)0733-9445(1988) 114:8(1804)

[20] Lai M, Ho J. Confinement effect of ring-confined concrete-filled-steel-tube columns under uni-axial load. Engineering Structures. 2014;67(5):123–141. DOI: 10.1016/j. engstruct.2014.02.013

[21] Krishan A L, Rimshin V I, Troshkina E A. Deformability of Volume-Compressed Concrete Core of Concrete Filled Steel Tube Columns. IOP Conf. Series: Materials Science and Engineering. 2020;753: 1-7. DOI: 10.1088/1757-899X/753/022053

[22] Ilyushin A A. Plastichnost. Moscow: Gostekhizdat; 1948. 376 s. (rus).

[23] Goode C D. Composite Columns – 1819 tests on concrete-filled steel tube columns compared with Eurocode 4. The Structural Engineer. 2008;86(16): 33-38.

[24] Leon R T, Hajjar J F. Limit State Response of Composite Columns and Beam-Columns. Part II. Application of Design Provisions for the 2005 AISC Specification. Engineereng Journal, AISC. 2008;45(1):21-46.

[25] Krishan A L et al. The Energy Integrity Resistance to the Destruction of the Long-Term Strength Concrete. Procedia Engineering. 2015;117(1): 211-217. DOI:10.1016/j. proeng.2015.08.143

Chapter 6

Concrete Performance in Cold Regions: Understanding Concrete's Resistance to Freezing/Thawing Cycles

Mohammed A. Abed and György L. Balázs

Abstract

This Chapter provides a detailed better understanding of the freeze/thaw effect on concrete, it is discussing the attack mechanism for both types of freeze/thaw deterioration: Internal frost damage and Surface scaling. Freeze/thaw attack is a serious problem for concrete but the most common physical deterioration type that shortening the life of concrete in cold environments. An Air-entraining agent is one of the solutions for reducing the effect of freeze/thaw cycles on concrete. Meanwhile Using supplementary cementitious materials in the production of concrete has different effects on the behavior of concrete exposed to freeze/thaw cycles. This chapter is discussing five of the common supplementary cementitious materials and their effect on concrete resistance to freeze/thaw cycles.

Keywords: freeze/thaw cycles, supplementary cementitious materials, internal frost damage, surface scaling

1. Introduction

The effect of freezing/thawing is one of the most common physical deteriorations of concrete in cold environments, which causes serious damages and induces cracks in concrete structures. This effect is clear in such European countries where the temperature drops below 0°C and ice start to form. The freezing of water inside the capillary pore structure causes an increase in the volume of approximately 9%, thus causing severe cracking and disruption of the concrete, especially if the pores in the concrete are close to saturation [1, 2]. Using high-quality and adequate amount of supplementary cementitious materials can enhance the concrete resistance to such issues. Thus, sustainable solutions can be useful not only for environmental aspects, but also for enhancing the concrete resistance to freeze/thaw cycles. Fly ash, silica fume, blast furnace slag, and Metakaolin are good examples for waste materials that enhance concrete's resistance to freeze/thaw cycles if used in appropriate amount.

2. Theoretical explanations of freeze/thaw attack

It was originally thought that the small, entrained air voids (deliberately introduced concrete to provide frost resistance) worked by providing sinks for the water

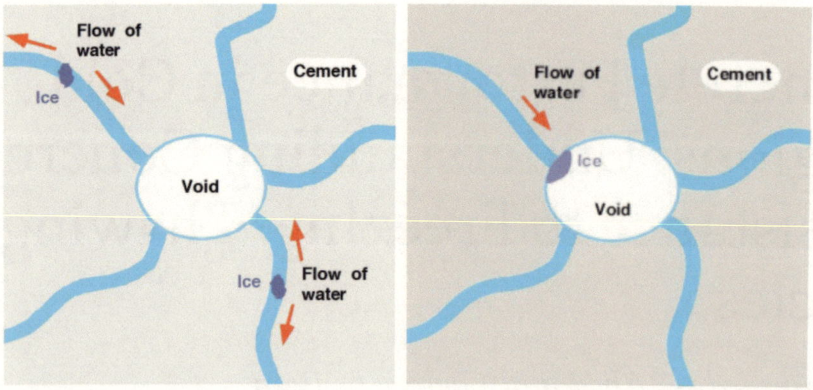

Figure 1.
Left (original concept): Air voids prevent excessive pore pressure by providing sinks for displaced water. Right (revised concept): Air voids provide a site for nucleation and growth of ice crystals drawing in liquid from pores [3].

displaced by the volume change associated with the transformation of water into ice, but it was subsequently realized that the voids serve as nucleation sites where ice crystals can grow without constraint. During the freezing process ice crystals in the air voids suck liquid from the small pores of the paste, this suction creates negative pressure in the pore liquid which puts the entire solid matrix into compression; thereby inducing compressive stress into the concrete, **Figure 1**. The shown figures have been presented at the concrete & cast stone conference in Boston, USA (2008) as well fib bulletin 53 [3].

Pore system characteristics plays an essential role in the transport properties of concrete as well as the behavior of concrete when it exposes to frost action. Capillary forces determine the absorbed water in concrete and by ice expanding hydraulic pressure increases and damage occurs. The process is directly proportional to the rate of temperature decrease, in addition, forming crystals of ice could interact with the walls of the capillary pores [4–6].

3. Types of freeze/thaw deteriorations

The cement paste combines several types of voids which directly affect its properties. The typical scales of both the voids and the solid phases in the hydrated cement paste are shown in **Figure 2**. While the hydration reaction progresses, the spaces that are initially filled with water, are replaced by the hydration product this residual space is called capillary pore [8]. According to **Table 1**, capillary pores are divided into three groups: small, medium, and large capillary pores.

Since the pores which are smaller than 10 μm in diameter have less influence on permeability, capillary pores are defined as medium and large capillaries, with diameters from 10 nm to 10 μm. The pores in this range would mostly affect the permeability and diffusivity of the cement paste. The hydrated product occupies more volume than the cement particles, and with the development of the hydration reaction, in the gel form continues to expand into the capillary system. Physical deterioration and damage-inducing processes causing cracking and other effects in concrete structures can arise from various causes including freeze/thaw effects. Concrete structures are periodically exposed to the deteriorating effect of freezing/thawing damage and that comes in two types [7, 11, 12]: internal frost damage and surface scaling. The former is caused by the freezing water inside the

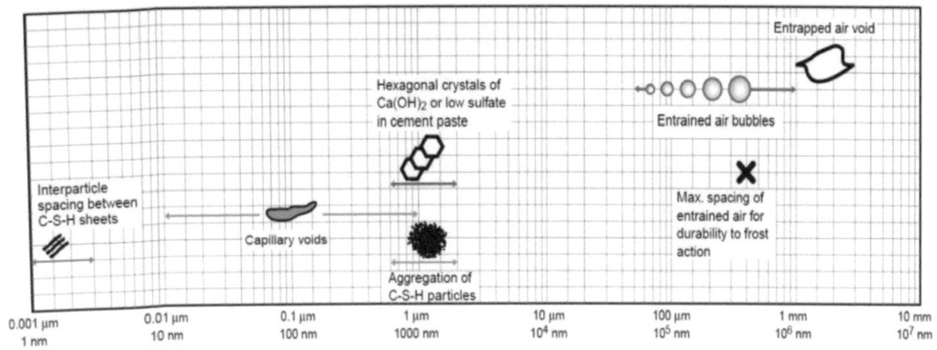

Figure 2.
Typical dimensions of different phases in the hardened cement paste [7].

Name	Type of pores	Diameter	Paste properties affected
Micro pores "inter layer"	Gel pores	Up to 0.5 nm	Shrinkage, creep at all RH
Micro pores "inter layer"		0.5 nm to 2.5 nm	Shrinkage, creep at all RH
Small (gel) capillaries		2.5 nm to 10 nm	Shrinkage between 50% and 80% RH
Medium capillaries	Capillary pores	10 nm to 50 nm	Strength, permeability, shrinkage at high RH, >80%
Large capillaries		50 nm to 10 μm	Strength, permeability
Entrained air		0.01 mm to 1 mm	Strength

Table 1.
Classification of pores in hydrated cement paste [9, 10].

concrete body, as a result of internal frost damage; weight change and compressive strength loss can occur. When the concrete surface comes into contact with weak saline solutions, surface scaling occurs and causes small flakes or chips of concrete on the surface.

The two types of freezing/thawing damages are visualized in **Figure 3**. The adoption of air-entraining admixtures is one of the best solutions for enhancing the freeze/thaw resistance of concrete [11]; the addition of a specific amount of appropriately sized air voids allows for the accommodation of any increase in water volume in case of freezing. As shown in **Figure 4**, air-entraining agent enhances the concrete's resistance to freeze/thaw cycles, as sample A (with air-entraining) experience less deterioration after 300 freeze/thaw cycles than sample B (without air-entraining) [14].

However, the result of such treatment is not always satisfactory, and the debate is ongoing. Thus, standards have recommended limitations regarding air void parameters, such as the spacing factor between voids and minimum air content of the fresh mixture [15, 16]. The recommended level of air-entraining is about 5%, which already causes a reduction in mechanical properties. It is necessary to compensate for this loss by technological steps to retain the required class of concrete.

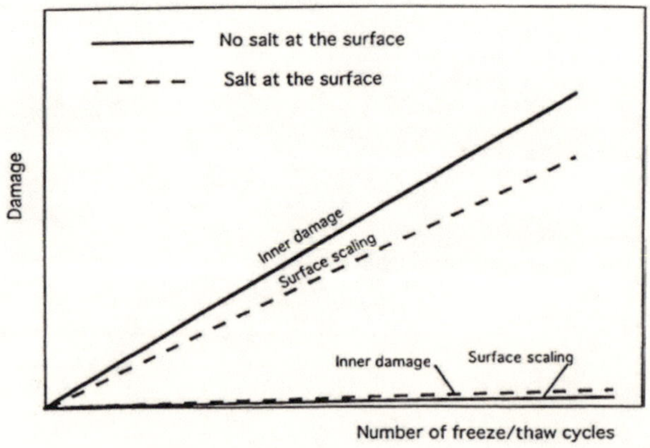

Figure 3.
The two main types of internal frost damage [13].

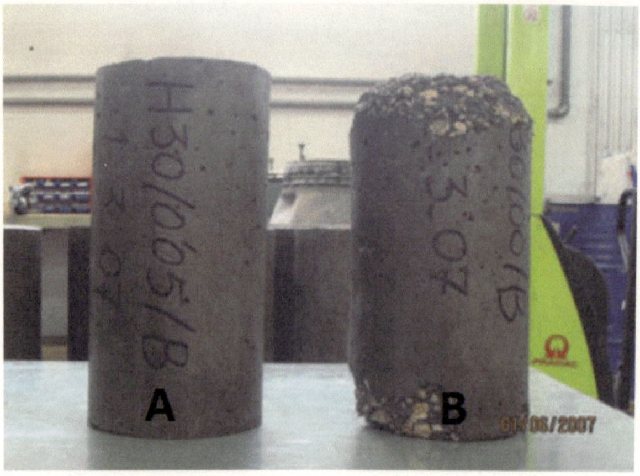

Figure 4.
Specimens A and B after 300 freeze/thaw cycles, where B without air-entraining agent but A with an appropriate air-entraining agent (0.2) [14].

4. Factors that affected the freeze/thaw resistance

Susceptibility to freeze/thaw damage is affected by the concrete's composition, permeability, porosity, type, moisture content, age, air-entraining admixtures, workability, exposure environment, and aggregate type. Where it observed that the key factors relevant to freeze/thaw durability are the air void characteristics especially in the case of using admixtures, they should compatible, in concrete and high quality of aggregate. As well as mix design limitations according to BS EN 206:2013 + A1:2016 [17] are to be considered necessary [18].

Yet not sufficient conditions for achieving high frost resistance of concrete, especially when exposed to severe frost, wet conditions, and high levels of de-icing agents. Thus, following the principles transpiring from the literature's observations during the design of freeze/thaw resistance concrete can be highly beneficial. Some of these observations have been concluded as [11]:

- High-quality aggregate should be used featuring appropriate frost resistance tested in NaCl solution and water, high resistance to fragmentation, continuous grading, and minimal water requirement.

- When specifying admixtures like Supplementary cementing materials, to concrete it is necessary to check their compatibility by valuating the air void characteristics in hardened concrete, and this concerns in particular air-entraining and water-reducing admixtures.

The severity of freezing (minimum temperature); rate of fall of air temperature (rate of freezing); change of air temperature during periods of freezing; the number of freeze/thaw cycles and the presence of de-icing salts is the major environmental aspects and critical factors that are participated in the developing of freeze/thaw damage.

5. Evaluation of the freeze/thaw deterioration

There are many test methods by which the internal damage and surface scaling of concrete can be evaluated. ASTM C 666 [19] and ASTM C 672 [20] are the typical tests by which the internal damage of the concrete specimen and the scaling of the specimen surface in the presence of de-icer salts can be assessed prospectively. However, numerous national freeze/thaw tests resemble the previous direct tests performed by using actual freezing and thawing loads. In the following two sections the used procedure for testing the internal frost damage and surface scaling test is explained.

5.1 Internal frost damage

The following procedure is usually used to apply the internal frost damage test:

- Submerging concrete cubes in water until full saturation.
- Lifting reference concrete cubes in water and placing the rest in a laboratory freezer until a specific number of freeze/thaw cycles.

  According to CEN/TR 15177 [21] each freeze/thaw cycle includes (two hours of cooling, two hours of freezing at - 20°C, two hours of thawing, and 2 hours at + 20°C). Concrete cubes are surrounded by air in the freezing phase, while surrounded by water in thawing phase.

- Lifting the concrete cubes in water for three days after specified number of freeze/thaw cycles, which are determined based on the purpose of the study. This is to prevent any residual freezing from occurring inside the concrete cubes.
- Finally, testing the residual compressive strength and/or determining the weight change.

Figure 5 shows the time–temperature curve in the centre of the concrete sample, where: 1 / freeze/thaw cycle; 2 / temperature range in the reference; Y / temperature in °C; and X / time in h.

5.2 Surface scaling

In accordance with CEN/TS 12390–9 [22] (slab test) surface scaling is evaluated. The main idea the test is to determine the scaled materials amount after exposing the concrete surface to 56 freeze/thaw cycles, where the tested concrete surface should be in contact with a 3% NaCl solution. The testing procedure begins by

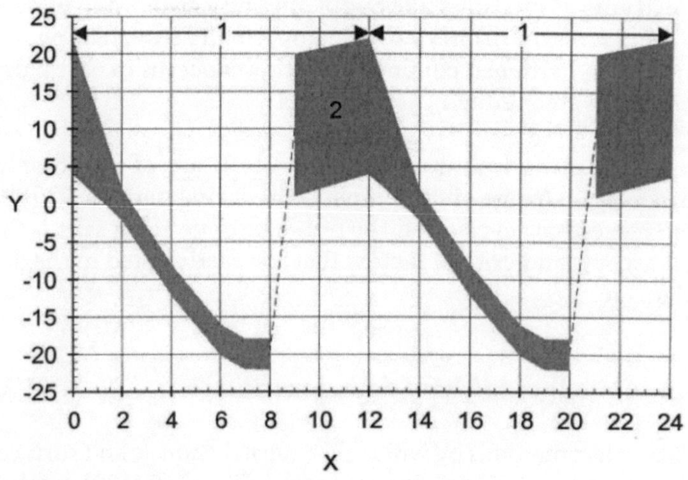

Figure 5.
The time–temperature curve in the Centre of the concrete sample [21].

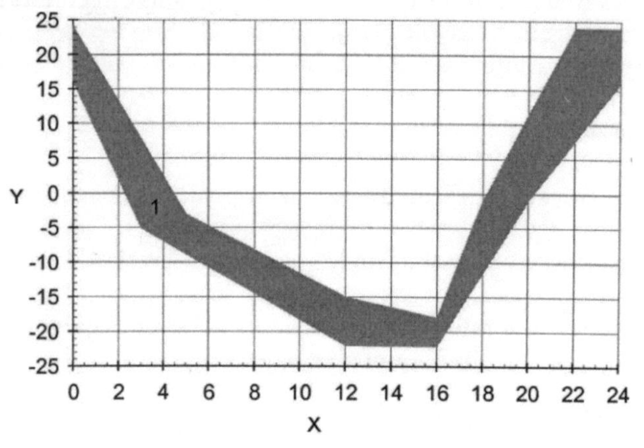

Figure 6.
Time–temperature curve in the freezing medium at the Centre of the test surface [22].

cutting concrete cubes in half and using the sawn surface to come into contact with 3% NaCl solution, while the other surfaces must be isolated. For each specimen (half concrete cube), 5 mm thickness of 3% NaCl solution should be placed in contact with the sawn surface sand placed in freezer to start the freeze/thaw cycles. Each freeze/thaw cycle includes (six hours of cooling, six hours of freezing at −20°C, six hours of thawing, and six hours at +20°C). After a specified number of freeze/thaw cycles, up to 56 cycles, the scaled material is collected from the tested surface and weighed, the higher the weight, the higher the scaling.

Figure 6 shows the time–temperature curve in the freezing medium at the centre of the test surface, where: 1 / temperature range at the centre of the test surface, whereby the time of temperature > 0°C is 7 to 9 h; Y / temperature in °C; and X / time in h.

6. Internal frost damage mechanisms

The durability of concrete refers to its ability to withstand deterioration due to harsh environmental conditions. These conditions can act alone or together and include heating and cooling, freezing and thawing, wetting and drying, chemical attacks, and abrasion. Deterioration due to freezing/thawing causes D-cracking to occur [23]. Under harsh service conditions, the durability of reinforced concrete structures is related to concrete frost resistance. Frost resistance tests are accompanied by the accumulation of residual dilation deformations affected by temperature-humidity stresses, ice formation and other factors. It is affected directly by the porosity, which is an integral part of the concrete structure which is formed as a result of cement hydration [24]. If the porous material is so wet that the theory of hydraulic over-pressure governs the freezing phenomenon, pore water is squeezed into the larger air-filled pores and the external environment surrounding the sample and causes there an abrupt increase of relative humidity. However, if the pore system is not filled with pore water to the extent that hydraulic pressures are induced into the material then after the first freezing of the pore water an under-pressure is formed in the pore system and the sample contracts [25], **Figure 7**.

D-cracking is a type of freeze/thaw damage in concrete pavements, it occurs due to the poor-quality coarse aggregates. By increasing the wet level of coarse aggregate reaching saturation, it becomes more susceptible to damage during freezing/thawing cycles. Pressure builds up inside of the coarse aggregate as a result of water freezing inside its pores. If the pressure due to the expansion of the water within the pores of the coarse aggregate is higher than its internal strength, the coarse aggregate will crack. Then the deterioration process accelerated due to the increased potential water availability, where the interfacial transition zone between the coarse aggregate and cement matrix is supposed to be slightly thicker. Its porosity can be 100% and that increases and accelerates the freeze/thaw deterioration [23, 26, 27].

Sicat et al. [28], investigated experimentally the real-time deformational behavior of the interfacial transition zone in concrete during freeze/thaw cycles. They observed that due to its high porosity and weak strength of the interfacial transition zone, its deformation is higher than that of cement matrix and aggregate. The deformation has been experimentally proven by taking a closer look at the unbroken sections of the interfacial transition zone of the specimens by electron microscope after the freeze/thaw cycles in **Figure 8**. This deformation is clearer in the case of wet specimens as well as much significant in higher water to cement ratios.

It is observed that the deterioration due to freeze/thaw cycles increases by increasing the number of cycles, and sometimes a certain number of freeze/thaw cycles is required before the deterioration occurs, and that could be the effect

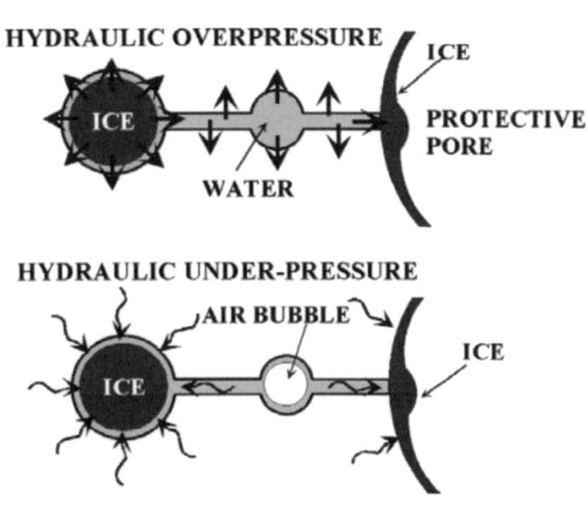

Figure 7.
Water transfer mechanisms in the freezing theories of wet porous materials [25].

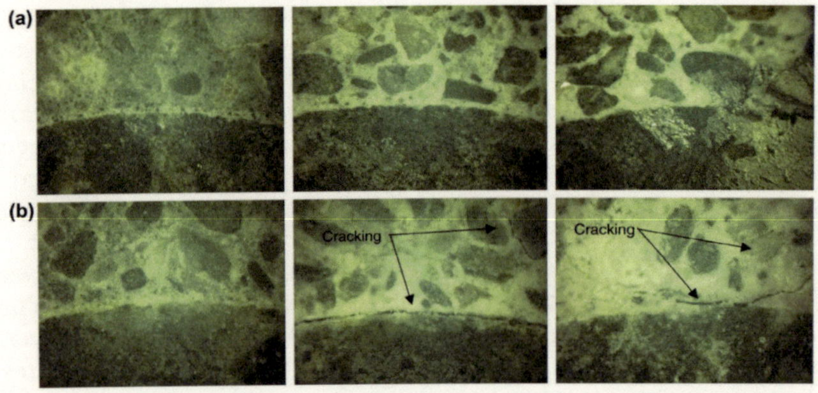

Figure 8.
Interfacial transition zone after freeze/thaw cycles (from left to right 30%, 50%, and 70% w/c); (a) dry conditions, (b) saturated conditions [28].

of fatigue. As each freeze/thaw cycle is added to cumulative internal deterioration this is similar to the normal mechanical fatigue with constant load cycles. However, it considers as low-cycle fatigue, where normally the freeze/thaw cycles are less than 300.

The reason behind the fatigue hypothesis is that one often notices a development of damage of the type shown in **Figure 9** when the material is tested in so-called "open" freeze/thaw, i.e. in a test where the specimen has access to water, during freezing and/or during thawing. In some cases -curves D, P, C in **Figure 9** damage increases progressively with the number of freeze/thaw cycles already from the first cycle, in other cases, all other curves a certain number of freeze/thaw cycles are needed to initiate.

The residual compressive strength of concrete after the effect of the freeze/thaw cycle is tested by Lu et al. [30] as shown in **Figure 10, Left**. It shows the changing of compressive strength after 0, 25, 50, and 75, freeze/thaw cycles respectively when the strain rate remained constant. They observed that the compressive strength decreased linearly with an increasing number of freeze/thaw cycles, where after

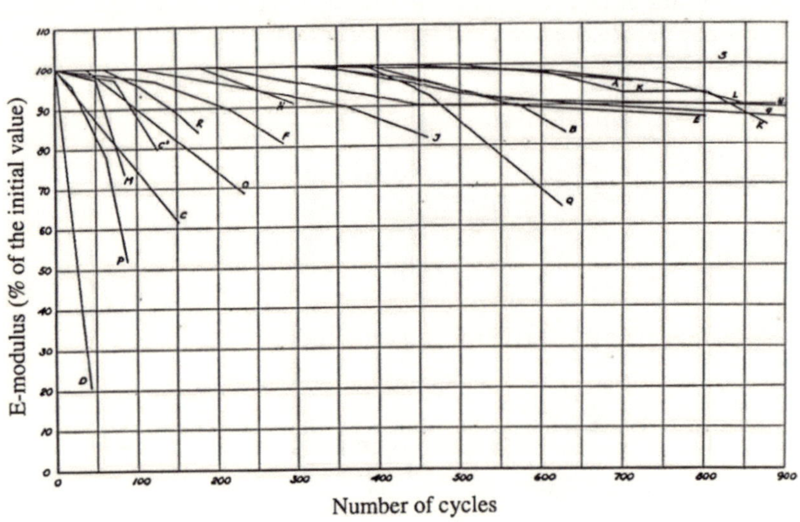

Figure 9.
Reduction in E-modulus of cement mortar specimens repeatedly frozen in the air to −15°C and thawed at water +5°C, 2 cycles per day [29].

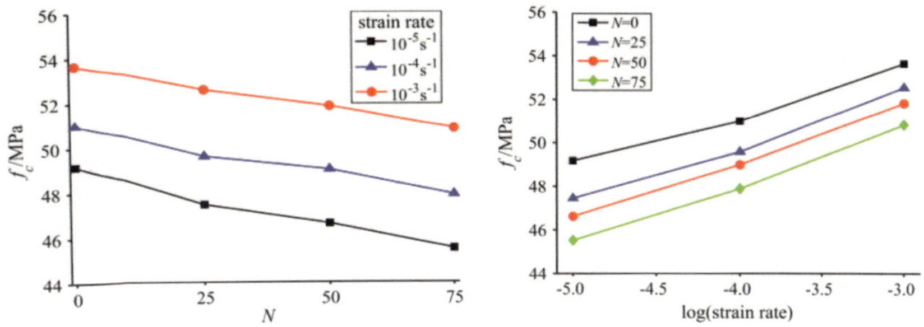

Figure 10.
Left: Compressive strength versus freeze/thaw cycles. Right: Compressive strength versus strain rates [30].

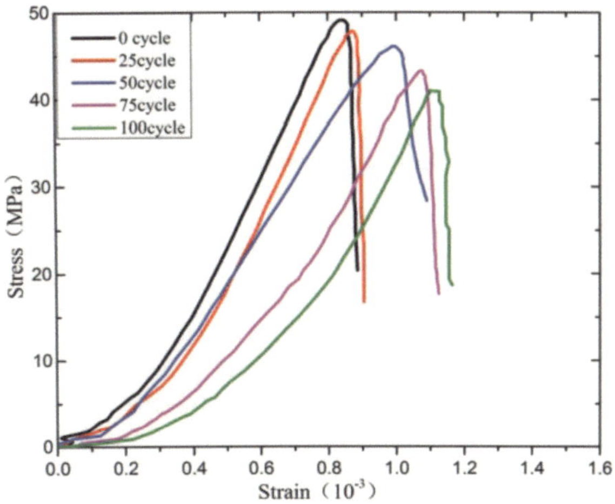

Figure 11.
Typical stress–strain curves for concrete at strain rate 1.0 × 10–2/s under different freeze/ thaw cycles [31].

repeated freeze/thaw cycles, the cracks in the concrete will pass through each other, and their strength will gradually decrease, and finally even completely lost. However, the compressive strength increases with strain rate in a nearly linear progression as shown in **Figure 10, Right**. Under freeze/thaw cycles, crack propagation becomes faster with a higher stress rate, however, enhancement of concrete strength could happen due to the coarse aggregate that stops the crack extension.

They find a decrease in the residual compressive strength of concrete if the concrete was subjected to a fatigue compression loading prior to freeze/thaw cycling and this reduction was increased by increasing the fatigue cycles. Fatigue cycles cause microcracks and that increase in the irreversible tensile strain due to freeze/thaw cycles afterwards. Han and Tian [31], presented the stress–strain relationships after different freeze/thaw cycles as plotted in **Figure 11**. They concluded that the concrete deformation experienced four deformation stages: compaction, elastic deformation, plastic deformation, and finally fracture. The peak load decreases as the freeze/thaw cycling increase where microcracks occurs as a result of freeze/thaw cycling, its number affects the microcrack number and width.

7. Mechanism of action of the air-entraining admixtures

For adequate resistance of concrete to the freezing/thawing cycles; air voids should not have a spacing factor (the maximum distance between two air voids in concrete) larger than 0.2 mm or 0.008 in. For a given air content, the size of the air voids cannot be too large if the proper spacing factor is to be achieved. The specific surface should be greater than 24 mm2/mm3 [32], where the air void system consists mainly of three parameters: the void's size, the space factor, and the void size distribution [33, 34]. As previously mentioned, to produce freeze/thaw resistant concrete; a proper air-void system is required in the concrete to accommodate the volume extension of water when it freezes without causing damages.

Artificially, the use of effective air-entraining admixtures can ensure the stabilization of the air-void system and produce air-entrained concrete which is one of the greatest advancements in concrete technology [35]. The term "air-entrainment" refers to the air deliberately introduced into concrete by adding air-entraining admixture. In the case of using air-entraining admixtures, the entrained air voids reduce the hydraulic pressure by behaving as expansion chambers despite the volume increment when water turns into ice [36, 37]. The schematic mechanism of the performance of concrete (with and without air-entraining admixture) exposed to freezing is shown in **Figure 12**. Šelih [39] and Wang et al. [40], investigated the freeze/thaw resistance of concrete with air-entraining admixture and low water content and they experienced a better performance of 10 times lower mass loss comparing to the normal one.

Air-entraining admixture is either surfactant that reduces the surface tension of water, or substances that produces a water-repellent precipitate which is required to produce the air voids dispersed throughout the concrete and that ultimately provides durability in freezing/thawing situations. Surfactant air-entraining admixture secured the best overall air void characteristics, followed by salt-type air-entraining admixture containing tall oil, and then, salt type air-entraining admixture containing Vinsol resin and wood rosin. However, some types of air-entraining admixtures, including vinsol resin, sodium adipate, sodium oleate, do not reduce the surface tension of water [16, 41]. The fresh state of concrete seriously affected its freeze/thaw resistance like the flowability and setting time of concrete. Air voids have a lifetime and should not be unstable, where they could be collapsed due to different fundamental physical mechanisms. Such as the diffusion of air from a void to a larger one, the void coalescence due to capillary flow, or the rapid hydrodynamic drainage of liquid between voids [33].

Other types of chemical admixtures are usually used to control the flowability and viscosity of concrete such as superplasticizers and viscosity modifying admixtures, which also affect the hardened properties of concrete and generally they affected the

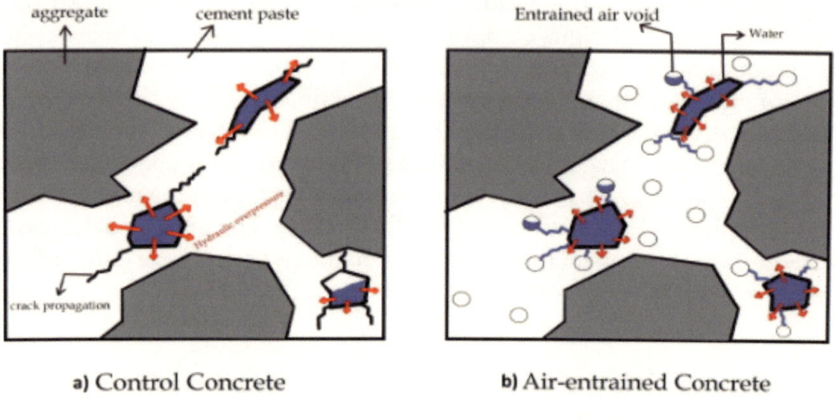

a) Control Concrete   b) Air-entrained Concrete

Figure 12.
Performance of normal and air-entrained concrete exposed to freezing a) cumulative hydraulic pressure oversaturated voids b) mitigating hydraulic overpressure condition by the distribution of air bubbles [38].

dosage rate of air-entraining admixtures. Viscosity modifying admixture increases the mix viscosity and its mode of action depends on the type and concentration of the polymer in use, while the superplasticizer causes a reduction in total air void surface areas and increases in air void spacing factors [31, 42]. Meanwhile, polycarboxylate superplasticizers usually have an air-entraining effect, wherewith the use of polycarboxylate superplasticizers; the air voids characterize with smaller diameters than voids formed as a result of lingosulphonicor naphthalene superplasticizers [41, 43].

The usage of admixtures such as viscosity modifying admixtures and superplasticizers can reduce the ability of an air-entraining admixture to create a proper air void system. Where the air content seems to be decreased with the increase of viscosity modifying admixture content and that will probably necessitate greater additions of air-entraining admixture to secure a given air volume [44, 45]. However, and generally; because of the complexity of modern air-entraining admixtures and other chemical admixtures, it is impossible to generalize the effects of their interactions with surfactants on the air entrainment.

8. Effect of supplementary cementitious materials on freeze/thaw resistance

Recently, the use of supplementary cementitious materials has dramatically increased due to an increase in environmental awareness. Incorporation of various mineral admixtures, supplementary cementitious materials, into concrete as one of the cement-based composites is a generally well known and frequently used approach. Supplementary cementitious materials are usually employed for partial replacement of cement in the production of concrete, where their use brings significant ecological and economic benefits [12].

Fly Ash, blast furnace slag, silica fume and metakaolin are commonly used supplementary cementitious materials for the purpose of enhancing the performance of concrete from a different point of view, such as enhancing the concrete's properties, value, and cost. They have a clear effect on the fresh, mechanical and durability performance of concrete, and thus on its Freeze/thaw resistance. Several studies conducted on the effect of supplementary cementitious materials on concrete performance have been reviewed [46–48]. Most of these studies have indicated that adding supplementary cementitious materials helps in resisting deleterious effects, such as alkali-silica reactivity, freeze/thaw deterioration, random cracking, and permeability. Supplementary cementitious materials enhance the frost resistance through reducing the macro capillary porosity of cement matrix and form stable gel-like hydration products [24].

8.1 Cement type

Concrete is rich in alkali and therefore it reacts actively with acidic gasses and liquids, freeze/thaw resistance of hardened cement paste plays an important role in the resistance of concrete to the freeze/thaw cycles. Especially in the case of using high-quality coarse aggregate with low porosity. The freeze/thaw resistance of cement paste depends on its porosity, the size of pores, capillaries, and their distribution [49]. Skripkiūnas et al. [50], tested the freeze/thaw resistance of concrete made with different types of cement, the effect of four types of cement on the freeze/thaw resistance of concrete has been investigated by them: CEM I type Portland cement, CEM II/A-S 42.5 N and CEM II/A-LL 42.5 R contained 6–20% of blast furnace and limestone, and CEM III/B 32.5 N-LH contained 66–80% of slag. They concluded that concrete containing slag cement CEM III/B 32.5 N-LH has the highest freeze/

thaw and de-icing salt scaling resistance. While concrete containing Portland cement CEM I 42.5 R has the lowest freeze/thaw and de-icing salt scaling resistance.

The ductility of concrete made of cementitious composites decreased remarkably and the cement containing slag is freeze/thaw resistant [51]. Deja [52] observed a high salt scaling resistance of concrete containing cement rich in granulated blast furnace slag and by using air-entraining admixture even at relatively high values of w/c ratio. Skripkiūnas et al. [53] investigated hardened cement paste made of CEM I 42.5 R cement modified with synthetic zeolite admixture, the mass loss and deformations freezing/thawing cycles were much lower in concrete modified with synthetic zeolite which modifies the morphology of hardened cement paste. However, in other research for Skripkiūnas et al. [54], they investigated the freeze/thaw resistance of concrete made of CEM I 42.5 R cement modified with sodium silicate solution and found that the destruction after 56 freeze–thaw cycles and exposure do de-icing salt solutions is smaller in hardened cement paste modified with sodium silicate solution. It may be used to improve the durability of hardened cement paste and concrete used in road building.

8.2 Fly ash

Fly ash is one of the most popular mineral additives that used in the concrete industry, it helps in improving the consistency of fresh concrete and reducing the hydration heat. The use of fly ash has been investigated extensively in the literature and it is found that it is beneficial for the durability of concrete, and it enhances the permeability properties. Replacement cement by over 40% of fly ash causes a serious reduction in the service life. Fly ash has a low carbon content and very variable properties, thus its effect on freeze/thaw resistance is not clear. Despite of the complexity of defining the influence of fly ash on freeze/thaw resistance, it is favorable due to its positive effect on the long-term durability of concrete [47, 48, 55].

8.3 Blast furnace slag

Blast furnace slag is a by-product formed during pig iron production. The incorporation of ground blast furnace slag into concrete is very beneficial because of its positive effect on the durability properties [56]. Blast furnace slag has a positive effect on many engineering properties, such as bleeding, consistency of the fresh mixture, the heat of hydration and permeability. The relatively high bulk density of a cement matrix modified by slag frequently leads to a decrease in the frost resistance, however, the results of experimental studies are inconsistent [57]. This could be partly a result of the different experimental procedures and also because of the use of modifying agents such as plasticizers, which significantly affect the character of the porous system.

8.4 Silica fume

Using silica fume as a substitution of cement usually does not exceed 5–10%. Therefore, its effect on the frost resistance seems to be marginal, provided the air content is maintained on a level, that is high enough. Besides, the long-term tests indicate, that a negative effect occurs only when the amount of silica fume exceeds 10% [13], **Figure 13**.

8.5 Metakaolin

Metakaolin has been generally used as a filler or supplementary cementitious material in concrete, and it was proved to be able to enhance its performance.

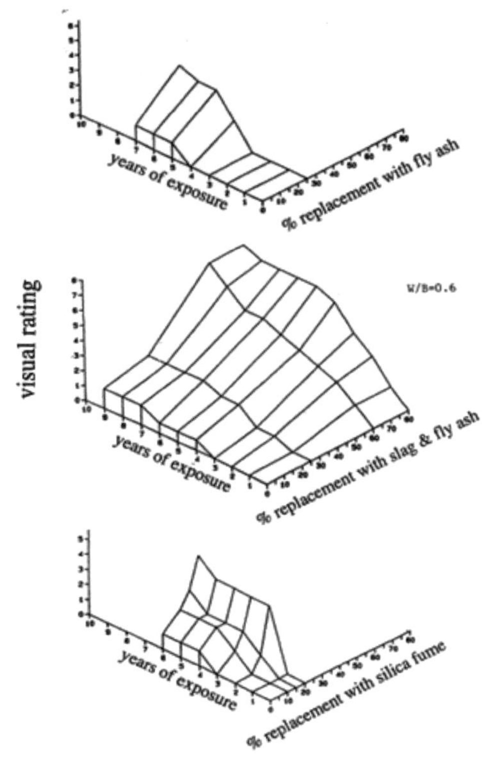

Figure 13.
Result of exposure tests on the long-term behavior of air-entrained concrete containing fly ash, blast furnace slag and silica fume [13].

The introduction of metakaolin into concrete as a supplementary cementitious material can improve its freeze/thaw resistance significantly. Where metakaolin decreases the total porosity and improves the pore structure of concrete which is directly related to the mechanical properties [58]. It was found that the freeze/thaw resistance corresponds to the content of insoluble hydrates in the cement paste. However, the used binder system also significantly affects the character of the concrete capillary system. In general, modern concrete mixtures with reduced water to cement ratio exhibit increased frost resistance; nevertheless, this approach tends to increase the autogenous shrinkage, which could cause propagation of surface cracks during hardening in the case of a poor curing regime, resulting in the reduction of the durability and especially the frost resistance [51, 55].

9. Conclusions

This chapter discusses theoretically the freeze/thaw attack of concrete in terms of its mechanism affected factors. Two types of freeze/thaw attack were discussed, Internal frost damage and Surface scaling. Either type of freeze/thaw attack, increasing the number of freeze/thaw cycles leads to deteriorating the properties of concrete, owning that to the weakened bond between aggregates and paste caused by the development of internal cracks in the cement paste with repeated cycles. Using high-quality aggregate and/or supplementary cementitious materials can enhance the concrete resistance to such issues. However, checking the compatibility of the used materials in terms of evaluating the air void characteristics is mandatory in particularly when the air-entraining agent is used. Fly Ash, blast furnace slag, silica fume, and metakaolin are commonly used SCMs for the purpose of enhancing the performance of concrete. It is concluded that adding supplementary cementitious materials helps in resisting deleterious effects, such as alkali-silica reactivity, freeze/thaw deterioration, random cracking, and permeability.

Acknowledgment

Authors acknowledge the support by the Hungarian Research Grant NVKP_16-1-2016-0019 "Development of concrete products with improved resistance to chemical corrosion, fire or freeze-thaw".

Conflict of interest

No conflict of interest.

Author details

Mohammed A. Abed[1*] and György L. Balázs[2]

1 Rutgers University, New Jersey, United States of America

2 Budapest University of Technology and Economics, Budapest, Hungary

*Address all correspondence to: mohammed.abed@rutgers.edu

References

[1] Coussy, O. & Monteiro, P. J. M. 2009. Errata to "Poroelastic model for concrete exposed to freezing temperatures" [Cement and Concrete Research 38 (2008) 40-48. Cement and Concrete Research, 39, 371-372.

[2] Zeng, Q., Fen-Chong, T., Dangla, P. & Li, K. 2011. A study of freezing behavior of cementitious materials by poromechanical approach. International Journal of Solids and Structures, 48, 3267-3273.

[3] fib 2009. Structural Concrete Textbook on behaviour, design and performance, Second edition Volume3: Design of durable concrete structures, International Federation for Structural Concrete.

[4] Nili, M., Azarioon, A. & Hosseinian, S. M. 2017. Novel Internal-Deterioration Model of Concrete Exposed to Freeze-Thaw Cycles. Journal of Materials in Civil Engineering, 29, 04017132.

[5] Sun, Z. & Scherer, G. W. 2010. Effect of air voids on salt scaling and internal freezing. Cement and Concrete Research, 40, 260-270.

[6] Liu, L., Shen, D., Chen, H., Sun, W., Qian, Z., Zhao, H. & Jiang, J. 2014. Analysis of damage development in cement paste due to ice nucleation at different temperatures. Cement and Concrete Composites, 53, 1-9.

[7] Mehta, P. & Paulo, J. M. M. 2006. Concrete: Microstructure, Properties, and Materials, McGraw-Hill Education.

[8] Neville, A. M. & Brooks, J. J. B. 2010. Concrete Technology, Pearson Education.

[9] Mindess, S., Young, J. F. & Darwin, D. 2003. Concrete, Prentice Hall, Pearson Education, Inc. Upper Saddle River, NJ 07458, U.S.A.

[10] Ying, W. 2013. Performance Assessment of Cement-Based Materials Blended with Micronized Sand: Microstructure, Durability and Sustainability. PhD, Delft University of Technology.

[11] Glinicki, M. A., Jaskulski, R. & Dąbrowski, M. 2016. Design principles and testing of internal frost resistance of concrete for road structures - critical review. 2016, 15, 23.

[12] Abed, M. & Nemes, R. 2019. Long-term durability of self-compacting high-performance concrete produced with waste materials. Construction and Building Materials, 212, 350-361.

[13] Göran, F. 1995. Freeze-thaw resistance of concrete : destruction mechanisms, concrete technology, test methods, quality control : a contribution to the BRITE/EURAM project BREU-CT92-0591 "The Residual Service Life of Concrete Structures", Stockholm, Division of Building Materials, LTH, Lund University.

[14] Al-Assadi, G., Casati, M. J., ´Andez, J. F. & ´Alvez J. C. G. 2010. Effect of the curing conditions of concrete on the behaviour under freeze–thaw cycles. Fatigue & Fracture of Engineering Materials & Structures, 34, 461-469.

[15] ACI. 2008. Guide to Durable Concrete, ACI Committee Reports, American Concrete Institute.

[16] Łaźniewska-Piekarczyk, B. 2013. The frost resistance versus air voids parameters of high performance self compacting concrete modified by non-air-entrained admixtures. Construction and Building Materials, 48, 1209-1220.

[17] BS EN 206:2013+A1:2016. 2013. Concrete. Specification, performance, production and conformity.

[18] Abed, M. 2019. Green Self-compacting High-performance Concrete PhD, Budapest University of Technology and Economics

[19] ASTM C666 / C666M-15. 2015. Standard Test Method for Resistance of Concrete to Rapid Freezing and Thawing. West Conshohocken: ASTM International.

[20] ASTM C672 / C672-12. 2012. Standard Test Method for Scaling Resistance of Concrete Surfaces Exposed to Deicing Chemicals. West Conshohocken: ASTM International.

[21] CEN/TR 15177. 2006. Testing the freeze-thaw resistance of concrete — Internal structural damage.

[22] CEN/TS 12390-9. 2016. Testing hardened concrete. Freeze-thaw resistance with de-icing salts, Scaling.

[23] Koubaa, A. & Snyder, M. B. 2001. Assessing Frost Resistance of Concrete Aggregates in Minnesota. Journal of Cold Regions Engineering, 15, 187-210.

[24] Trofimov, B. Y., Kramar, L. Y. & Schuldyakov, K. V. 2017. On Deterioration Mechanism of Concrete Exposed to Freeze-Thaw Cycles. IOP Conference Series: Materials Science and Engineering, 262, 012019.

[25] Penttala, V. 2006. Surface and internal deterioration of concrete due to saline and non-saline freeze–thaw loads. Cement and Concrete Research, 36, 921-928.

[26] Kyle, R., Brett, B., Amir, M. & Heather, M. 2013. Effects of Curing Methods and Supplementary Cementitious Material Use on Freeze Thaw Durability of Concrete Containing D-Cracking Aggregates. A cooperative transportation research program between Kansas Department of Transportation, Kansas State University Transportation Center, and The University of Kansas.

[27] Mohammed A. Abed Indirect Evaluation of the Compressive Strength of Recycled Aggregate Concrete at Long Ages and after Exposure to Freezing or Elevated Temperatures. Russ J Nondestruct Test 57, 195-202 (2021). https://doi.org/10.1134/S1061830921030025

[28] Sicat, E., Gong, F., Ueda, T. & Zhang, D. 2014. Experimental investigation of the deformational behavior of the interfacial transition zone (ITZ) in concrete during freezing and thawing cycles. Construction and Building Materials, 65, 122-131.

[29] Göran, F. 2000. Fatigue effects associated with freeze-thaw of materials. Report TVBM (Intern 7000-rapport). Sweden: Lund University.

[30] Lu, J., Zhu, K., Tian, L. & Guo, L. 2017. Dynamic compressive strength of concrete damaged by fatigue loading and freeze-thaw cycling. Construction and Building Materials, 152, 847-855.

[31] Han, N. & Tian, W. 2018. Experimental study on the dynamic mechanical properties of concrete under freeze–thaw cycles. Structural Concrete, 19, 1353-1362.

[32] Ley, M.T., Chancey, R., Juenger, M.C.G. & Folliard, K.J. 2009. The physical and chemical characteristics of the shell of air-entrained bubbles in cement paste. Cement and Concrete Research 39, 417-425.

[33] Du, L. & Folliard, K.J. 2005. Mechanisms of air entrainment in concrete. Cement and Concrete Research 35, 1463-1471.

[34] Jin, S., Zhang, J. & Huang, B. 2013. Fractal analysis of effect of air void on freeze–thaw resistance of concrete. Construction and Building Materials 47, 126-130.

[35] Collins, A.R. 1944. The Destruction of Concrete by Frost. Journal of the Institution of Civil Engineers 23, 29-41.

[36] Shang, H., Cao, W. & Wang, B. 2014. Effect of Fast Freeze-Thaw Cycles on Mechanical Properties of Ordinary-Air-Entrained Concrete. The Scientific World Journal 2014, 923032.

[37] Shang, H.-S. & Yi, T.-H. 2013. Freeze-Thaw Durability of Air-Entrained Concrete. The Scientific World Journal 2013, 650791.

[38] Ebrahimi, K., Daiezadeh, M.J., Zakertabrizi, M., Zahmatkesh, F. & Habibnejad Korayem, A., 2018. A review of the impact of micro- and nanoparticles on freeze-thaw durability of hardened concrete: Mechanism perspective. Construction and Building Materials 186, 1105-1113.

[39] Šelih, J. 2010. Performance of concrete exposed to freezing and thawing in different saline environments. Journal of Civil Engineering and Management 16, 306-311.

[40] Wang, K., Nelsen, D.E. & Nixon, W.A. 2006. Damaging effects of deicing chemicals on concrete materials. Cement and Concrete Composites 28, 173-188.

[41] Chatterji, S. 2003. Freezing of air-entrained cement-based materials and specific actions of air-entraining agents. Cement and Concrete Composites 25, 759-765.

[42] Łaźniewska-Piekarczyk, B. 2013. The type of air-entraining and viscosity modifying admixtures and porosity and frost durability of high performance self-compacting concrete. Construction and Building Materials 40, 659-671.

[43] Litvan, G.G. 1983. Air Entrainment in the Presence of Superplasticizers. Journal of the American Concrete Institute 80, 326-331.

[44] Khayat, K.H. 2000. Optimization and performance of air-entrained, self-consolidating concrete. ACI Structural Journal 97, 526-535.

[45] Lachemi, M., Hossain, K.M.A., Lambros, V., Nkinamubanzi, P.-C. & Bouzoubaâ, N. 2004. Self-consolidating concrete incorporating new viscosity modifying admixtures. Cement and Concrete Research 34, 917-926.

[46] Hanna, K., Morcous, G. & Tadros, M. K. 2014. Effect of Supplementary Cementitious Materials on the Performance of Concrete Pavement. Journal of Materials in Civil Engineering, 26, 789-793.

[47] Wang, D., Zhou, X., Meng, Y. & Chen, Z. 2017. Durability of concrete containing fly ash and silica fume against combined freezing-thawing and sulfate attack. Construction and Building Materials, 147, 398-406.

[48] Yazıcı, H. 2008. The effect of silica fume and high-volume Class C fly ash on mechanical properties, chloride penetration and freeze–thaw resistance of self-compacting concrete. Construction and Building Materials, 22, 456-462.

[49] Kumar, R. & Bhattacharjee, B. 2003. Porosity, pore size distribution and in situ strength of concrete. Cement and Concrete Research 33, 155-164.

[50] Skripkiūnas, G., Nagrockienė, D., Girskas, G., Vaičienė, M. & Baranauskaitė, E. 2013. The Cement Type Effect on Freeze – Thaw and Deicing Salt Resistance of Concrete. Procedia Engineering 57, 1045-1051.

[51] Özbay, E., Karahan, O., Lachemi, M., Hossain, K. M. A. & Atis, C. D. 2013. Dual effectiveness of freezing–thawing

and sulfate attack on high-volume slag-incorporated ECC. Composites Part B: Engineering, 45, 1384-1390.

[52] Deja, J. 2003. Freezing and de-icing salt resistance of blast furnace slag concretes. Cement and Concrete Composites 25, 357-361.

[53] Skripkiūnas, G., Nagrockienė, D., Girskas, G. & Janavičius, E. 2012. Resistance of modified hardened cement paste to frost and de-icing salts, The Baltic Journal of Road and Bridge Engineering 7(4), pp. 269-276.

[54] Skripkiūnas, G., Girskas, G. & Nagrockienė, D. 2012. Freeze-thaw and de-icing salt resistance of the hardened cement paste modified with synthetic zeolite, Proc. of the International Congress on Durability of Concrete (ICDC 2012), 18-21 June, 2012, Trondheim, Norway, pp. 1-13.

[55] Reiterman, P., Holčapek, O., Zobal, O. & Keppert, M. 2019. Freeze-Thaw Resistance of Cement Screed with Various Supplementary Cementitious Materials. REVIEWS ON ADVANCED MATERIALS SCIENCE.

[56] Özbay, E., Erdemir, M. & Durmuş, H. İ. 2016. Utilization and efficiency of ground granulated blast furnace slag on concrete properties – A review. Construction and Building Materials, 105, 423-434.

[57] Giergiczny, Z., Glinicki, M. A., Sokołowski, M. & Zielinski, M. 2009. Air void system and frost-salt scaling of concrete containing slag-blended cement. Construction and Building Materials, 23, 2451-2456.

[58] Gyurkó, Z., Szijártó, A. & Nemes, R. 2019. Cellular concrete waste as an economical alternative to traditional supplementary cementitious materials. Journal of Thermal Analysis and Calorimetry, 138, 947-961.

Section 2

Production and Applications of Supplementary Concrete

Chapter 7

Production of Sustainable Concrete by Using Challenging Environmentally Friendly Materials Instead of Cement

Abebe Demissew Gashahun

Abstract

The environmental problems accompanying concrete come from cement. This means that the final product i.e., concrete is an environmentally sociable material by itself. This guides us to play on the concrete constituents which cause the largest environmental impact, which is cement. Therefore, if we can abate cement amount and increase cementing materials which can substitute cement for concrete, we will be able to minimize the concrete impact on the environment. The saving of cement quantity in concrete can be realized by substituting it with diverse extra cementitious materials which are a by-product of another industry and waste of agriculture.

Keywords: Cement, Concrete, Cementous Materials, Sustainable

1. Introduction

The industry of construction is one of the booming economic sectors on the planet now a day. The sector comprehends constructions of horizontal and vertical structural elements. In Ethiopia, it was happened the same, since the governmental policy support infrastructure development projects to realize the transformation to the industry from agricultures. Due to this an exploitation of naturally deposited resources for concrete imputes becomes the main agenda for the government advanced than the other. Beyond this, concrete is concrete a blend of aggregates that id either crushed stone or gravel or sand or blending of them and cement past which mixture of cement, water, chemical admixture, and cementitious materials. The paste contains cement and water and sometimes other cementitious and chemical admixtures, whereas the aggregates contain sand and gravel or crushed stone. The cement paste in the concrete helps to make strong bonds between aggregate particles. Aggregates, major ingredients of concrete by volume, are comparatively inactive fillers material that has a share of about 68–85% of concrete and can therefore be expected to influence its properties. The cement which is among the main concrete ingredient plays an excellent role but is that the costliest and environmentally inimical material.

Since that cement production requires high consumption energy and leads to discharges of greenhouse gas, there continues to be a global search for new binders and admixtures that could partially replace traditional ordinary hydraulic cement

IntechOpen

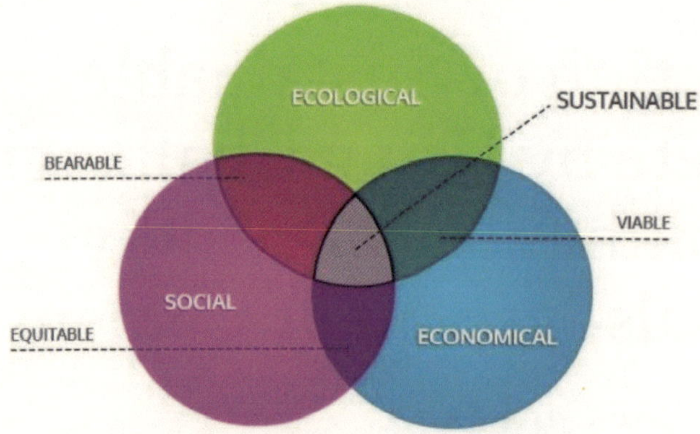

Figure 1.
Inter-relations among the ecological, economic, and social impact of construction.

and improve the environmental sustainability and sturdiness of concrete structures. The application of left-over by-products in construction materials as replacement of concrete constitutes become an attractive alternative to disposal and an eco-friendly solution to the challenges concerning the exploitation and shortage of non-renewable natural resources in the globe.

Sustainability is demarcated as a combination of environment, economy, and society, and among this parameter for development with sustainability, the environment is the dominant one due to deterioration of our environment is driving the current worldwide focus on sustainable development. Generally, almost all scholars agreed definitions of sustainability are "Meeting the desires of the present generation without compromising the ability of future generations to meet their needs." As shown in **Figure 1**, to have sustainable construction outputs, there should be a balance among environmental (ecological), social and economic aspects of building (construction) activities.

2. Construction industry

The term construction is mostly accustomed designate the physical (tangible) infrastructure and related facilities, every type of activity associated with the erection and repair of immobile structures & facilities. Construction is creating or assembling of something from something and it produces a one-of-a-kind product, it's complex and undertaken through cooperation by a short-lived organization, and it's essential to the growth of a given country and a crucial sector within the nation's economy (in Ethiopia, about 60% of the federal capital budget is channeled to this sector for Physical infrastructures, 70% of the capital budget is allotted for Transport & Communication and Buildings cover only 13% of the capital budget). The sector has a significant and dynamic part in transforming the aspirations and desires of individuals into reality by physically implementing various construction development projects. The products of construction contribute extensively towards the creation of wealth and consequently the quality of the lifetime of the population.

The construction business furnishes capital improvements to countries, which is incredibly much allied with the event of investments to supply future benefits to nations. Since the industry primarily represents an investment, construction activities drop quite other industries during recessions. The construction industry often makes skills more immediately rewarding which is why most workers during this industry became more prosperous professionals than in other industries. The industry is pigeonholed by converging (many components and lots of flows going together into one object), temporary (project being funded only for one object), highly fragmented with significant negative impacts, more complex & often associated with changes & uncertainties, amplified reaction in financial conditions, labor-intensive and land dependence. Outside this, it's challenged by low productivity, cost and time target failure, quality failures (increase construction costs in rework alone), conflicts and disputes, leading to claims and time-consuming litigation, waste during construction (Unnecessary material handling & material waste, rework, poor material allocation, lack of constructability, etc.), image (it is a sector having the worst public image among industrial sectors, Dirty, dull, & environmentally insensitive), have the very best incident rate of fatal accidents and somber injuries of all industries and also the furthermost environmentally unfriendly industries.

Beyond its role in the nation's development, the sector incorporates a negative bearing on the built-up and its surrounding milieu or sustainable development. The industry affects sustainability in its major five phases: (1) pre-design phase (material selection, building program, project budget, team selection, partnering, project schedule, codes, standards, laws, research, and site selection); (2) on-site phase (site analysis & assessment, site layout and development, watershed conservation and management, equipment and materials on-site); (3) design phase (passive solar design, materials & specification, indoor air quality): (4) construction phase (environmentally conscious construction, preservation of features & vegetation waste management and source control practices): (5) operation and maintenances phase (maintenance plans, indoor quality, energy efficiency, resource efficiency, renovation, housekeeping & custodial practices).

3. Concrete productions

Concrete construction demands land and hence the sourcing of materials for construction activities (aggregates, cement, waters, admixtures, etc.) from quarry sites and borrows pits can potentially end in the whole removal of vegetation and virgin materials. Additionally, the displacement of individual, concrete ingredients production resalted for losing of important ecological resources for local people, vegetation that gives watershed protection, and as a result the diminution of biodiversity of national or regional or global importance.

The aggregate manufacturing steps have many considerable environmental impacts. The foremost obvious environmental effect resulted from stone, aggregates, and mineral mines for industries is the degraded quality of air, and related health effects, sourcing from airborne emission from both stack and also disturbed areas at these mines. Natural deposits sources of aggregates are being depleted and causing a heavy threat to the environment and likewise to society. Because of a high rate of natural aggregates depletion from its source beds causing lots of problems such as loss of strata of water-retaining, bank slides, exposing water supply scheme intake wells, dropping underground water table levels which become a cause for agricultural effect and aquatic life disturbances.

In emerging nations like Ethiopia, due to rapid urbanization and infrastructure projects, there is a wide expansion of cement industries that releasing this pollutant to the surrounding. Apart from these environmental concerns regarding CO_2 emission during cement manufacturing, natural resource-demanding also makes cement expensive when compared with aggregates and water for concrete productions. Consequently, to overcome these problems, searching for more environmentally friendly and economical materials that have cement properties has prolonged attention in other such types of materials that can be used fully or partially substitute normal Portland cement.

4. Cementiouse materials

Cement is a material with cohesive and adhesive characteristics that make it capable of bonding mineral fragments into a compact whole and having major roles in concrete for the construction industry. But in the reveres, it is not environmental-friendly and the most expensive concreting materials.

Usage of concrete mix design with optimum cement content, enhancement of concrete durability, and use of supplementary cementing materials are the focus areas for sustainability in concrete industries.

Therefore, requirements for durable, economical, and more environmental-friendly ingredients for concrete, particularly for cement, have stretched curiosity to other cementing construction materials which can be used as partially or fully replaced the normal Portland cement for the sustainable construction industry.

The cement which is one of the basic ingredients of concrete is the fine gray powder and when it reacted with water it forms to harden, rigid and stable structures which bonded aggregates together acting like glue and gives the desired strength of concrete. The first invention of cement started when Romans, mixed lime ($CaCO_3$) with volcanic ash, producing cement mortar which was used during the construction of monumental structures as coliseum [1]. In accumulation to this, cement is defined as a mineral chemical produced by mixing a well-defined ratio of raw materials at highly elevated temperatures. Cement is a universally known and applicable construction material throughout the world. Besides, the depletion of inputs raw material for manufacturing of cement is one of the environmental effects, there is plenty of emissions of CO_2 to the atmosphere. It is believed that one tone of cement clinker production creates almost an equivalent ton of CO_2 and other greenhouse gases [2]. This implies that the quantity of cement produced is directly proportional to gas emission to the environment. And also, it shows that the cement factories tremendously contribute to today's worldwide apprehension, which is global warming. Furthermore, to its releases of different gases, extraction of raw materials is environmentally unfriendly due to degradation and disturbance of the existing natural environment. This shows that the cement industry contributes to today's worldwide concern, which is global warming. This endangers the sustainability of the cement factory and that of concrete. Beyond this cement industry needs high capital investment, energy-intensive, and highly dependent on power and transport and it leads cement as the greatest environmentally unfriendly and costliest imputes of concrete for the following three reasons.

a. Cost of production

The cement factory is one of the most energy-consuming/intensive industry in the globe and consequently, 30–40% of the total production cost it goes to fuel and energy cost for production. The cost of raw materials represents the second-largest percentage of cement manufacturers' cost structures. The abundance of this ingredient for cement production is reliable in most parts of the country and the availability of these raw materials is justified by the distribution of cement factories throughout the country with its raw materials for cement factories. However, variances across regions and companies depend on the operating efficiency of each producer comparative to others.

b. Ingredients for cement production

The major ingredients for the production of cement include clay, limestone, marl, chalk, and others, noteworthy quantities of which are endlessly quarried to service the demand for cement. Substitute materials have been sourced to substitute for traditional natural ingredients. The cement sector at present uses huge quantities of power station fly ash, blast furnace slag, natural pozzolana, limestone, and silica fume, mainly to substitute for natural raw materials in the production process of blended cement such as pozzolana fly ash and granular ground bluest farness. The use of these alternative materials has significant economic benefits and positive environmental advantages. The needs to quarry primary raw materials are reduced, energy consumption in cement production is cut, and overall reductions in emission of dust, CO_2 and acid gases are attained. In some applications, the performance of concrete can be enhanced when these alternative materials complement Portland cement clinker.

c. Energy/power

Cement production is one of the energy-intensive processes. The specific thermal energy-demanding of a cement kiln varies between 3,000 and 7,500 million joules for a ton of clinker, depending on the basic process design of the plant. The explicit electrical energy-demanding typically ranges between 90 and 130 kWh and 60–130 Kg of fuel oil per ton of cement." The industry of cement was expected to produce 4.7 million tons per year to meet the demand in 2015, 27 million tons per year. However, the industry achieved an output of only 11.17 million tons of it in the year 2009/2010″. This result suggests the need to increase the production and supply capacity of cement to meet the need of the fast-growing construction industry.

4.1 Pozzolanic materials

The material of pozzolan is stated as an aluminosilicate/siliceous materials which are finely ground style and chemically react with calcium hydroxide within the presence of moisture it creates calcium silicate hydrate (CSH) and other cementitious materials. Clay and shale, opalinc chert, volcanic ash, and diatomaceous earth are an example of natural Pozzolanas while fly ash, rice husk ash, blast furnace slag, coffee husk ashy, silica fume, bagasse ash, and metakaolin are an example of artificial Pozzolanas. Most pozzolans used today are mainly widely available by-product materials. Since pozzolan has a variety of diversity, its chemical structure and contents also vary. Therefore, classifying Pozzolanas only depending on their chemical composition would be difficult. For this reason, ASTM C-618 classifies Pozzolanas depending on a performance basis as tabulated in **Table 1** [3–6].

The reason behind using Pozzolanas is the improvement found on both the hardened and fresh state concrete. Lowering of the thermal shrinkage and heat of hydration, increase in water tightness, decrease in the alkali-aggregate reaction, resistance to sulfate attack, better workability, and price effectiveness are some of the improvements achieved by using Pozzolanas blended with cement [7–9].

Partially or cement replacing materials are special construction materials either naturally occurring materials or industrial wastes/byproducts or agricultural wastes which could be castoff for concrete production. And they rely on activation of by-products while incorporating minimal amounts of cement are promising low carbon candidates that can potentially complement the globe's growing concrete industries by using the equivalent performance concept.

Chemicals	Pozzolan Class	
	F	C
$SiO_2 + Al_2O_3 + Fe_2O_3$ (min %)	70	50
MgO (max %)	...	5
SO_3 (max %)	5	5
Moisture content (max %)	3	3
Loss on Ignitions (max %)	12	10
Available alkalis as Na_2O (max %)	1.5	1.5

Table 1.
Chemical requirement for Pozzolanic materials.

Due to the increase of awareness of environmental concerns and natural resource consumptions, the issue of energy saving has been gradually emphasizing by the public. Owning to the considerable use of concrete and cement material, the natural material resources associated with the construction sector have been continuously reducing in recent years. However, for each country particularly for developing countries like Ethiopia, concrete is the most significant material for fundamental and public constructions. Thus, an innovative and alternative concrete material, which possesses feasibility and practicality, is critical and significant for mitigating environmental impact and promoting energy-saving performance. For this purpose, the most communal and practice in real concrete production for the cement replacement are natural pozzolana, Diatomaceous Earth, Glass Residue, Silpoz, fly ash, Corn Cob Ash, Ground granulated blast furnace slag, Silica Fume, Highly reactive metakaolin, rice husk ash (RHA), Bagasse ash, Coffee husk ash, calcined termite hell, water hyacinth ashy, etc.

i. Natural pozzolans

Natural pozzolans originated from volcanic activities are worldwide available materials, with varied compositions and subsequently a varied performance. However, for the reason that huge content in amorphous silica, it is usually an excellent material to be used as cement replacement. Curiously, before the invention of ordinary Portland cement, volcanic ash and air lime mixtures were commonly used, with a good performance and proven durability. Difficulties in the usages of these products are lack of characterization and the varied composition of raw-material layers, sometimes within the same area. However, naturally occurring pozzolans are used successfully in cement composition and may be looked at with cement replacement potential with the replacing percentage of cement up to 20% by mass. The use of ash of volcano for concrete production helps to reduce chloride ion diffusivity of concrete, inhibiting the localized corrosion of steel and further concrete degradation. This addition also promotes lesser heat of hydration and higher setting time. The improved performance has been qualified to the refinement of the stomate structure and the pozzolanic action of volcanic ash. One of the probable weaknesses using of these natural pozzolana materials is their diversity. To minimize this problem, usually natural pozzolanic materials from different extraction heights are mixed before use [1–3, 6, 10].

ii. Diatomaceous earth

Diatomaceous earth, which is also acknowledged as diatomite/fossil flour, is a sedimentary material mainly constituted by diatom outer shells. This very fine powder formed by the external skeletons of these unicellular beings is extremely rich in silica, has high porosity and surface areas. It is usually commercialized after it has been subjected to the previous calcination to remove organic matter and its characteristics make it liable to be replacement materials of cement. Its application in concrete production is usually in a weight percentage of 10–20% over the cement binder weight.

iii. Glass residue

Traditional soda-lime glass, predominantly composed of silica, but with a high percentage of sodium and calcium is a common residue that is finely ground for posterior use and 20–30% of cement replacement without harmful effects, performed satisfactorily concerning alkali reactivity and drying shrinkage. In addition to this, glass powder can significantly reduce the chloride ion penetrability of the concrete.

iv. Silpozz

Silpozz is extracted from rice husk ash and finer than cement with a particle size of 25 micrometers which helps it to fill the gaps between the aggregate and cement i.e., the determinants of density and strength of concrete. Because of this, it reduces the cement amount in the given concrete proportioning, and consecutively it elevates the compressive strength of concrete by 10–20% and high resistance towards the chemical attack, abrasion, and reinforcement corrosion.

v. Corn cob ash

Demanding for corn cob ash as cement replacing alternative materials is increased due to less amount of organic content, which proved the binding properties of cement. When we use it as alternative cement replacing materials, it will serve up to 20% by mass, and having advantages increasing the water amount which help as to obtain the desired plasticity as well as the initial and final setting time. Probably it is due to the reduced cement surface area and hence the delayed hydration process. Corn cob ash may therefore be most applicable when a low rate of heat development is necessary.

vi. Fly ash

Fly ash is a cementitious supplementary material for concrete production concrete and a byproduct of the pulverized coal in electric power generating plants. And also, a material of fine-grained having alumina, silica, iron, and calcium as a major content and sulfur, sodium, magnesium, carbon, and potassium as a miner percentage and it will serve as a cement replacement by weight up to 15%.

vii. Ground granulated blast furnace slag

Blast-furnace slag is the iron manufacturing industry's byproduct having the necessary minerals calling as cementitious materials like aluminosilicates and silicates and calcium.

viii. Silica fume

Silica fume is a waste from the production of silicon/ferrosilicon alloy in an electric furnace from high-purity quartz with coal. It is used as cement replacing concreting materials in between 5–10% by mass and it is recommended to be used in high strength and impermeable concrete.

i. Highly Reactive Metakaolin

Highly reactive metakaolin has nowadays available as a highly active pozzolana concrete material. Contrasting slag, fly ash, or silica fume, it is not the byproduct rather manufactured from high-purity kaolin clay by calcination at temperatures in the region of 700 to 800°C. Unlike silica fume, which has above 85% SiO_2, and it contains equal proportions of SiO_2 and Al_2O_3 by mass.

ii. Bagasse ash

Bagasse is fiber from cellulose from the extraction sugar-bearing juice of sugarcane and has silica and alumina which are the most vital component of cement replacing materials. It is also finding in large amounts as a byproduct from factories of sugar.

iii. Rice husk ash

Rice husk ash is a byproduct of agriculture and used up to 20% to replace cement in concrete and it has a good tendency to reduce temperature for the production of high strength mass concrete.

iv. Coffee husk ash

The use of different replacing of cement materials has become a popular practice in the industry of construction. The chemical composition of coffee husk ash has a significant value of Al_2O_3 and SiO_2, which are major components of cement. It can replace cement about 10% and concrete produced from CHA has high potential as a source of environmental-friendly cementitious material that reduces pollution and provides a sound coffee waste management option.

v. Termite hill clay

The chemical properties of calcined termite hill clay powder include the fact that the material is pozzolanic with the sum of SiO_2 (38.82%), Al_2O_3 (23.98%), and Fe_2O_3 (11.68%) constituting 74.48% of the material, leading to the conclusion that termite hill clay powder calcined at the temperature of 650o C satisfied the requirement of ASTM C618 of a minimum of 70%. Moreover, classified as natural Pozzolana class N. It means, the material can produce the cementitious compound that has binding property upon reaction with calcium hydroxide gained from the hydration of cement. Therefore, calcined termite hill clay powder was found suitable to partially replace cement in the production of concrete. up to 11.3% by weight and it implies it can reduce the CO_2 emission by 11.3% due to cement production and it will save the natural materials by the same percentages.

vi. Water hyacinth Biochar

Water hyacinth Biochar, a carbonaceous solid material obtained through a pyrolysis process from solid waste materials, and has properties like extremely low thermal conductivity, high chemical stability, low flammability, ability to absorb water, and highly capture and store CO_2. Recently, these properties of biochar favor its use as a partial cement replacing material in concrete construction up to 5% of cement by weight. In addition to this, every tone of biochar used in a building's envelope means that the equivalent of more or less one tone of CO_2 is prevented from re-entering the atmosphere.

4.2 Advantages of cement replacing materials

4.2.1 *Environmental advantages*

The cement industry is an energy-intensive industry with energy typically accounting for about 40% of operational costs, i.e., excluding capital costs but including electricity costs. The production of cement involves the consumption of large quantities of raw materials, energy, and heat. Cement production also results in the release of a significant amount of solid waste materials and gaseous emissions. The cement manufacturing industry is under higher scrutiny these days because of the large volumes of CO_2 emitted. This industrial sector is thought to represent 5–7% of the total CO_2 anthropogenic emissions. Concern over the impact of anthropogenic carbon emissions on the global climate has increased in recent years due to growth in global warming awareness. In addition to the generation of CO_2, the cement manufacturing process produces millions of tons of the waste product cement kiln dust each year contributing to respiratory and pollution health risks. To produce 1 ton of clinker, the typical average consumption of raw materials is 1.52 tones.

The amount of clinker needed to produce a given amount of cement can be reduced by the use of supplementary cementitious materials such as coal fly ash, slag, and natural Pozzolanas (e.g., rice husk ash, coffee husk ash, and volcanic ashes). The addition of these materials into concrete not only reduces the amount of material landfilled (in the case of industrials byproducts) but also reduces the amount of clinker required per ton of cement produced. Therefore, replacing the portion of Portland cement with those cementitious materials can substantially reduce the environmental impact of concrete associated with cement production like consumption of raw materials and energy use as well as emissions to air and avoiding environmental pollution due to avoiding as a waste.

a. Energy saving

The cement industry plays a significant role in global energy consumptions. Worldwide the cement industry is one of the most energy-intensive sectors in which energy represents 40% of the total production cost. The energy consumption in cement manufacturing is mainly related to the production methods that are wet methods consume more energy than dry methods. For instance, in the dry method, 1450°C of temperature is needed for the production of clinker which accounts for 97.2% of the total and the remaining is for finishing and raw materials grinding with the share of 0.9% and 1.9% respectively.

b. Reduction of CO_2 emission

Sustainable development of the cement and construction industry about environmental impact is one of the biggest challenges. The production of one ton of Portland cement release approximately one ton of CO_2 into the atmosphere in the manufacturing process. The cement industry contributes about 5% of the total atmospheric CO_2 emissions globally. As a matter, of fact, we are now concerned by the environmental impact of civil engineering structures. Judicious use of those cementitious materials as a partial replacement of cement can result from a significant result reduction of the CO_2 footprint of concrete structures. Most of the CO_2 emissions and energy use in the cement industry are related to the production of the clinker; 63% of the CO_2 emitted during cement production comes from the calcination process, while the rest (37%) is produced during the combustion of fossil fuels to feed the calcination process.

c. Economic advantages

The production of cement is energy-intensive, depends on the availability of raw materials near the cement manufacturing area and natural disturbances due to the extraction of raw materials. The process is mainly classified into three, the raw material preparation process, the clinker burning process, and the finish grinding process. Of all these processes, clinker burning is the most energy-intensive process, accounting for about more than 97.3% of the fuel consumed and about 30% of the electric power consumption, and the rest about 40% of the electric power is consumed by the finish grinding process and about 30% by the raw material preparation. Fuel costs are a large part of the manufacturing cost of the cement industry, making cement plants have aggressive energy consumption. Moreover, the clinker burning process as shown above takes more than 97% of the fuel consumption, implying that it is the most expensive part of cement production.

5. Concrete and sustainability

Currently, Sustainability is an important issue all over the globe and it is affected by cement and concrete technology. The construction industry particularly cement and concrete are responsible for the production of 7% carbon dioxide of the total world CO2 emission. Green concrete capable of sustainable construction is characterized by the application of industrial wastes to reduce consumption of natural resources and energy and pollution of the environment. Replacement of materials over nominal concrete is what makes green concrete more environmentally friendly concrete.

Cement is a pillar to develop infrastructures in the given nations. At the same time, cement production affects the local environment and nearby communities. The environmental issues of cement manufacturing are related to local, regional, and global problems in their mining and mineral processing. The local problems include dust, ground subsidence, noise, vibrations, chemical contamination, tailings spills, scenic and local ecological degradation, and health problems among miners. Regional problems are acid rain and contamination of surface and/or groundwater spills of processing chemicals and stream sediment loading. Global problems are the effects of mineral use and anthropogenic greenhouse gases contributing to global warming. On the one hand, dust emission sources are kiln, crusher, grinders, clinker coolers, and material handling equipment, which are in crushing and pyro processing. Besides, clinker manufacturing of pyro-processing is a considerable source of emission such as Cement Kiln Dust (CKD), gases like CO_2, Sulfur oxide, nitrogen oxide, and dioxins [11–15].

Air emitted/vented from various stages of cement processing contains dust, SO_2, NOx, CO_2, and heavy metals which can negatively affect the air quality of the area. One of the most common methods for reducing the environmental effect of concrete is by adding recycled materials to the concrete. The increased environmental awareness and dwindling resources in conjunction with regulation-based impetus enforced by governments/regional councils of different countries have led to the research and development of products and processes that employ effective waste utilization.

6. Conclusion

The concrete industry is a major contributor to air pollution and the user of natural resources. As such it bears a special responsibility to make a contribution towards sustainable development that is commensurate with its size. It can do so by pursuing three goals: (1) Searching for cement production technologies that are less energy-intensive and cause less air pollution. Since such technologies will not be available in the foreseeable future, the more realistic approach is to reduce the need for Portland cement, primarily by increased use of supplementary cementitious materials, especially waste materials. (2) Replacing concrete ingredients with recycled materials, such as recycled concrete or waste glass. (3) Through careful concrete mix design and prudent choice of admixtures, improve the durability of structures such that they need to be replaced less frequently.

In addition to this, sustainable construction makes wise use of all the natural resources and a 50% reduction in energy use, improves occupant health, comfort, productivity, reduces pollution and landfill waste that is not easily quantified, a sustainable building may cost more upfront, but saves through lower operating costs over the life of the building, construction is designed as one system rather than a collection of stand-alone systems with the help of the integrated system approach. Saying all this, I recommend to researchers make material characterization on those potential cement replacing materials.

Conflict of interest

"The authors declare no conflict of interest."

Author details

Abebe Demissew Gashahun
Construction Technology and Management Academic Program, School of Civil and Water Resource Engineering, Debre Markos Institute of Technology, Debre Markos University, Debre Markos, Ethiopia

*Address all correspondence to: abebe_demissew@dmu.edu.et

References

[1] Peter Hewlett. Lea's Chemistry of Cement and Concrete. Elsevier Science & Technology; 2004.

[2] Naik T.R. and Moriconi G. Environmental-friendly durable concrete made with recycled materials for sustainable concrete construction. University of Wisconsin Milwaukee; 2006.

[3] Science E. Comparison of different waste materials used as cement replacement in concrete Comparison of different waste materials used as cement replacement in concrete 2019. https://doi.org/10.1088/1755-1315/357/1/012010.

[4] Gashahun AD. Assessment on Cement Production Practice and Potential Cement Replacing Materials in Ethiopia 2020;12:22-8.

[5] Akadiri PO, Chinyio EA, Olomolaiye PO. Design of A Sustainable Building: A Conceptual Framework for Implementing Sustainability in the Building Sector 2012:126-52. https://doi.org/10.3390/buildings2020126.

[6] Alconpat R. Use of supplementary cementitious materials (SCMs) in reinforced concrete systems – Benefits and limitations 2020;10:147-64.

[7] Demissew A, Fufa F, Assefa S. Partial Replacement of Cement by Coffee Husk Ash for C-25 Concrete, 2019;10:12-21.

[8] Marchetti E. Use of Agricultural Wastes as Supplementary Cementitious Materials 2020.

[9] Ramezanianpour AA. Cement Replacement Materials. n.d.

[10] Lahri A, Dixit S. Alternatives to Cement in Concrete – A Review 2015;6:50-6.

[11] Just H. How to Make Concrete More Sustainable 2015;13:147-54. https://doi.org/10.3151/jact.13.147.

[12] Struble L, Godfrey J. How Sustainable is Concrete ? n.d.:201-11.

[13] Cachim P, Velosa AL, Ferraz E. Substitution materials for sustainable concrete production in Portugal 2014. https://doi.org/10.1007/s12205-014-0201-3.

[14] Ioannou S. Alternative cementitious solutions for sustainable concrete in the Gulf region THE WET CAST 2019.

[15] Meyer C. Concrete and Sustainable Development 2002:1-12.

Chapter 8

Application of Supplementary Cementitious Materials in Precast Concrete Industry

Amin Akhnoukh

Abstract

Supplementary cementitious materials (SCMs) are increasingly incorporated into the concrete mix design. Silica fume, fly ash, and multi-wall carbon nanotubes are used to improve concrete mix properties. The objective of this chapter is to decipher the impact of different SCMs on the fresh and hardened concrete properties, including concrete flowing ability, initial strength, final strength, modulus of elasticity, and modulus of rupture. In addition, the impact of SCMs on mitigating the alkali-silica reactivity of concrete and increasing the hardened concrete long-term performance is investigated. Developed concrete mixes, incorporating SCMs, are used in fabricating different precast/prestressed bridge girders. The impact of improved concrete properties on precast girder performance in increased flexure, shear, and span-to-depth ratio significantly improves project sustainability and reduces the overall project life cycle cost.

Keywords: silica fume, fly ash, carbon nanotubes, high strength concrete, durability, alkali-silica reactivity, supplementary cementitious materials

1. Introduction

High-strength concrete is increasingly used in the construction market in the United States and on a global scale. High-strength concrete is characterized by high early strength, high final strength, increased durability, and improved long-term performance. The use of high-strength concrete is advantageous in heavy construction projects, especially in precast/prestressed bridge construction. The mechanical advantage of high-strength concrete is attributed to the incorporation of supplementary cementitious materials (SCMs) in partial replacement of Portland cement. SCMs with different sizes and fineness are incorporated in the stepwise replacement of cement to create a binary mix (using one SCM) or ternary mix (using two SCMs). Incorporated SCMs increase the amount of the binder in the mix, which increases the mix strength. In addition, the improved packing order of the cementitious matrix reduces the mix void ratio and, hence, improve its durability and long-term performance. High-strength concrete was primarily used in the design and construction of high-rise buildings in major European cities and large/crowded American States as New York, Illinois, and California. In addition, the high compressive strength enabled bridge design engineers to precast long-span bridge girders with very high span-to-depth ratios (span/depth ratio greater than 30).

Recently, concrete mixes with higher strength and improved long-term performance were developed using SCMs, steel fibers, and a very low water-to-powder ratio. These mixes, commercially available in the United States construction market under the term *ultra-high-performance concrete*, are standardized by different agencies including the Federal Highway Administration (FHWA) in the United States, Association Francaise de Genie Civil (AFGC) in France, and the Japanese Society of Civil Engineers (JSCE). In their definition of the UHPC, the aforementioned organizations define UHPC as a cement matrix with minimum compressive strength of 150 MPa due to the high proportion of SCMs and very low water-to-powder ratio, and significant tensile strength due to the incorporation of random high-strength steel fiber. Different UHPC proprietary mixes are available in the international markets with standard characteristics. Examples of the proprietary mixes are BSI "Beton Special Industrial" (special industry concrete) developed by Eiffage, Cemtec developed by LCPC, and different kinds of ductal concrete mixes jointly developed by Bouygues, Lafarge, and Rhodia. Ductal concrete marketed by Lafarge and Bouygues is the only proprietary UHPC mix commercially available in the United States local construction market.

This chapter introduces different types and classifications of concrete mix designs, based on mix strength, workability, and long-term performance, different types of SCMs currently used in developing special concrete mixes, their impact on mix properties, and the main impediments to the widespread of SCMs application in precast/prestressed concrete industry.

2. Special concrete mix development

The complexity of construction projects, and the need to buildings structures with increased heights, bridges with longer spans, and infrastructure projects with minimized maintenance necessitates the development of concrete mixes with superior characteristics, higher durability, and minimal need for maintenance. Thus, concrete mixes with increased rheology and high flowing ability, high early and final strength are required. In addition, increased concrete durability is desired to minimize the need for regular maintenance and reduce the project life cycle cost. Special types of concrete mixes are displayed in the following section.

2.1 Ultra-high-performance concrete

Ultra-high-performance concrete (UHPC), known in the European market as reactive powder concrete, is a new class of concrete developed in the 1990s. UHPC concrete are currently used in high-rise building construction in the United States, and in specific transportation, and defense applications. UHPC mixes are characterized by self-consolidation concrete (SCC) workability, fast setting, high early strength, and final compression capacity that reaches 250 MPa [1, 2]. UHPC is characterized by improved long-term performance, low voids, higher alkali-silica resistivity due to the incorporation of SCMs, and post-cracking stiffness due to the placement of random steel fibers within the mix [3, 4]. The aforementioned characteristics represent a combination of the advantages of high-performance, self-consolidation, and fiber reinforcement inclusion, as shown in **Figure 1**.

2.2 High-strength concrete

The term high-performance concrete (HPC) is used to describe concrete produced with selected high-quality mix constituents, optimized mix design, and low

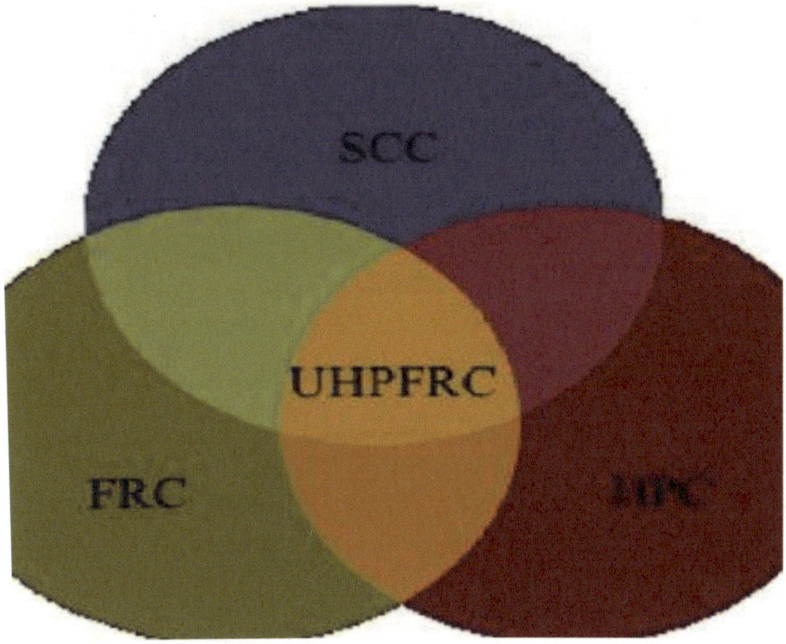

Figure 1.
Composition of UHPFRC [2].

water-to-powder (W/CM) ratio. According to the American Concrete Institute (ACI), HPC is defined as "concrete meeting special combination of characteristics and uniformity requirements," which cannot be achieved using conventional constituents, and regular mixing and curing procedures [5]. The mix composition of HPC depends mainly on the replacement of a significant amount of Portland cement and incorporating SCMs up to 40% by weight.

2.3 Self-X concrete mixes

Self-X concrete mixes are special types of mixes that provide specific advantage (s) to address a given project challenge. Examples of Self-X concrete mixes are as follows:

1. Self-healing concrete mixes, where developed hair cracks are treated internally using epoxy capsules incorporated in the mix,

2. Self-cleaning concrete mixes, known as photocatalytic concrete, used in maintaining the concrete surface texture and cleanliness by decomposing dirt and/or pollutants affecting the concrete, and

3. Self-consolidated concrete, known as self-compacting concrete (SCC), where the fresh concrete mix has no shear strength, and attains a water-like flowing ability. SCC mixes are usually used in pouring structural members with high depth and congested with heavy reinforcement as bridge girders.

Thus, SCC mixes usually incorporate silica fume for increased strength to sustain increased loading. Self-X concrete mix improved performance is shown in **Figure 2**.

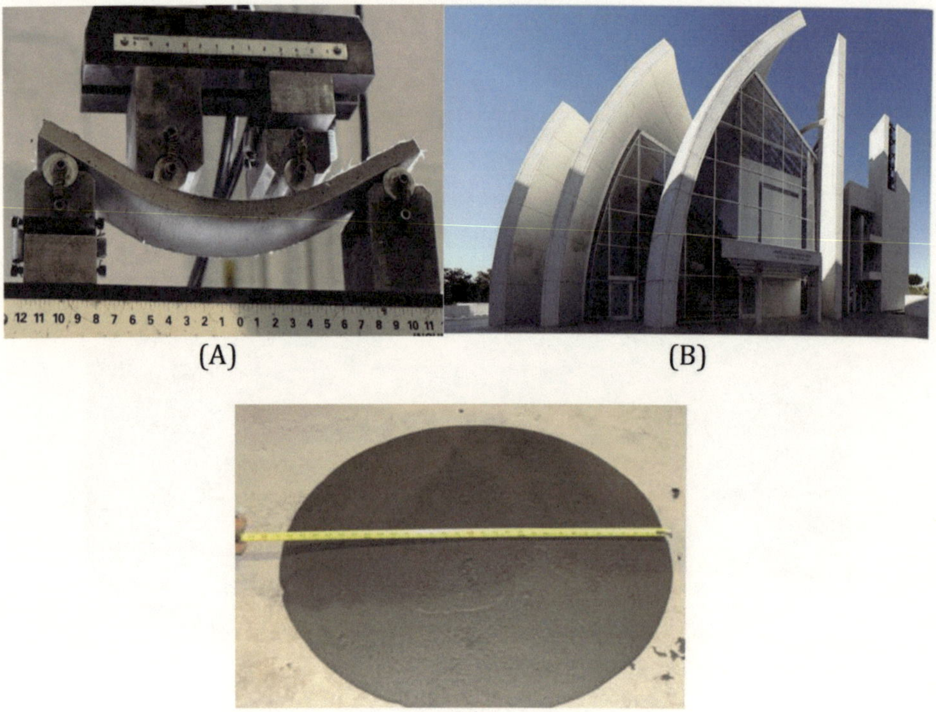

(A) (B) (C)

Figure 2.
(A) Self-healing concrete [6], (B) self-cleaning concrete [7], and (C) SCC concrete [8].

3. Supplementary cementitious materials (SCMs) in concrete industry

SCMs are increasingly used in concrete mix designs. Recently, binary and ternary concrete mixes are becoming the norm in heavy construction industries including the construction of bridges, tunnels, and culverts. Different types of SCMs are used according to the targeted characteristics of the developed mix. SCMs are available in granular shape with diameters ranging from nano-centimeters (as nano-silica) to a few millimeter diameter (as quartz flower). SCMs are used in partial replacement of cement. Cement weight up to 30% can be replaced by an equivalent volume of SCMs. Currently, different types of SCMs are available in the construction market including nano-silica, micro-silica, also known as silica fume, class C fly ash, class fly ash, quartz flour, blast furnace slag, and single and multi-walled carbon nanotubes. The characteristics of different SCMs, their impact on concrete mechanical properties, and their potential use are described in the following section.

3.1 Micro silica (silica fume)

Micro silica, also known as silica fume, is a byproduct of producing silicon metal or ferrosilicon alloys. Micro silica, as a mineral pozzolanic admixture, has a very fine particle size that averages 0.5 micro-meter in diameter. The average size of silica fume particle is 100 times finer than Portland cement size. Silica fume is commercially available in a densified and un-densified form, with similar chemical composition, and different densities. The chemical composition of silica fume is shown in **Table 1**.

Composition	SiO_2	Al_2O_3	Fe_2O_3	MgO	CaO	Na_2O	K_2O	SO_3
Percentage %	92–98	0.5	2.1	0.3	0.8	0.1	1.0	0.2

Table 1.
Chemical composition of micro-silica.

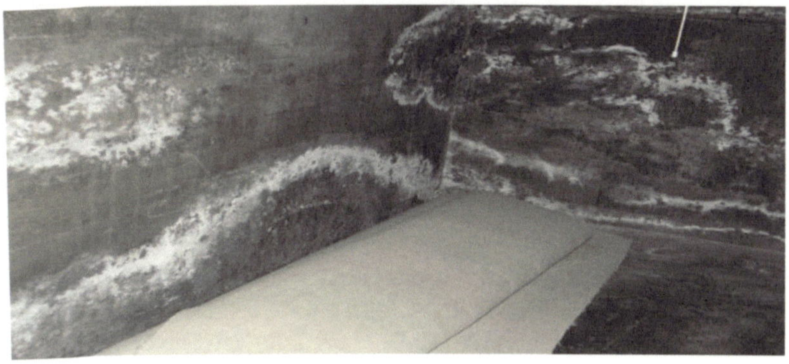

Figure 3.
Excessive surface salt due to efflorescence phenomena.

Micro-silica improves concrete mechanical properties through two different mechanisms by contributing to the Portland cement hydration. When mixing water is added to cement, without incorporating micro-silica, the following chemical reaction takes place during hydration:

$$\underset{}{\text{Cement}} + \underset{(\text{H2O})}{\text{Water}} \longrightarrow \underset{(\text{C}-\text{S}-\text{H})}{\text{Calcium Silicate Hydrate}} + \underset{\text{Ca(OH)}_2}{\text{Calcium Hydroxide}} \quad (1)$$

The main outcome of the hydration process includes the calcium silicate hydrate that acts as a binder and is directly responsible for the compressive strength of hardened concrete. The compressive strength depends on the amount of produced binder. Thus, the quantity of cement, cement fineness, water-cement ratio, and sufficient mixing energy and time are crucial to attain the required compressive strength of concrete upon hardening. The secondary outcome of the hydration process is calcium hydroxide, Ca $(OH)_2$, which does not act as a binder, and has no contribution to the strength of the mix. In addition, excessive amounts of $Ca(OH)_2$ may react with carbon dioxide and form a soluble salt that could leach within the hardened concrete causing efflorescence, as shown in **Figure 3**, and reduce the long-term performance of the structure due to its susceptibility to sulfate attacks, chemical attacks, and accelerating alkali-silica reactivity (ASR).

When micro silica is included in the mix, the added pozzolan reacts with the formed calcium hydroxide to produce an additional binder, which increases the hardened concrete strength and eliminates the salt formation by halting the efflorescence. The following equation describes the chemical contribution of micro-silica:

$$\underset{\text{Ca(OH)}_2}{\text{Calcium Hydroxide}} + \underset{(\text{SiO2})}{\text{Microsilica}} + \underset{(\text{H20})}{\text{Water}} \longrightarrow \underset{(\text{C}-\text{S}-\text{H})}{\text{Calcium Silicate Hydrate}} \quad (2)$$

In addition to its contribution to the concrete mechanical properties through the aforementioned reaction, micro-silica results in an improved packing order of the mixed granular material. The improved packing order results in reduced porosity and increased resistivity to chloride attacks and de-icing salts, and reduces the rate of steel reinforcement corrosion. Improved packing order of micro-silica concrete is shown in **Figure 4**.

3.2 Fly ash

Fly ash is a fine granular powder that exists in nature as a by-product of burning coal in power plants. Fly ash is used as a low-cost recycled material in concrete mixes to improve the concrete strength, reduce concrete viscosity and improve its pump ability, mitigate the ASR and its destructive impact on hardened concrete, and reduce the final cost of the produced concrete. The chemical composition of fly ash is shown in **Table 2**.

Two types of fly ash are used in the construction industry as SCMs to improve the fresh and hardened concrete properties. These are class C and class F fly ash. The main difference between the two types is that class F fly ash is a pozzolan; thus, it needs to react with the calcium hydroxide resulting from the cement hydration process to form the binding material. While class C fly ash is a cementitious material, it can produce binder directly once it reacts with water (direct hydration process).

The use of fly ash in concrete mixing provides fresh concrete with workability advantages. Fly ash particles are spherical in shape, which provides the concrete powder (cement, SCMs, and aggregates) with a higher tendency to flow. In addition to the lubricant effect provided, fly ash reduces the shear capacity of fresh concrete. Hence, the fly ash concrete mix has a higher tendency to flow, being pumped, and has a better hardened surface after formwork is removed.

Recent research shows that fly ash can partially replace up to 30% of the cement weight of conventional mixes. More than 30% fly ash could be used when mass concrete is poured [10]. The replacement of 20% of cement with an equivalent amount of fly ash results in improving the mix workability and/or maintaining similar workability while reducing 10% of mixing water.

In hardened concrete, class C fly ash provides improved mechanical properties due to its ability to form the additional binder. Class C and F fly ash results in reduced voids and lowered permeability in hardened concrete. Reduced concrete

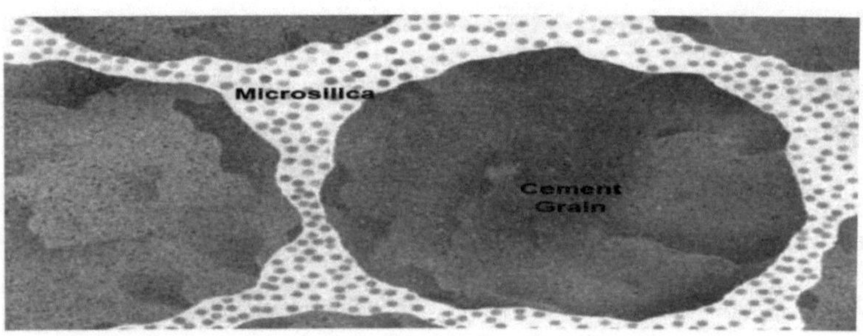

Figure 4.
Micro-silica concrete packing improved packing order [9].

Composition	SiO_2	Al_2O_3	Fe_2O_3	MgO	CaO	Na_2O	K_2O	SO_3
Percentage %	49	24.6	7.3	1.6	9.1	0.2	0.6	0.4

Table 2.
Chemical composition of fly ash.

voids play a major role in improving durability and blocking moisture dissipation to the hardened concrete, which improves its alkali-silica resistivity.

3.3 Quartz flour

Quartz flour, also referred to as crystalline silica, is a mineral used in UHPC and HPC mix development. Quartz flour is found in sand, granite, rocks, and some soil types. When these materials are cut, chipped, or drilled, dust evolves that contains the quartz flour particles. The diameter of quartz flour particles enables quartz flour to fill the voids within cement particles, which reduces the mix permeability and improves the resistivity of concrete to adverse environmental conditions. Quartz flour is used as a supplementary cementitious component in proprietary UHPC mixes produced by Lafarge and is commercially available in global markets under the name Ductal. Proprietary UHPC mix composition including quartz flour is shown in **Table 3**.

3.4 Blast furnace slag

Blast furnace slag is produced by grinding the glassy granular by-product of the steel industry. The fine ground blast furnace by-product is highly cementitious and contains calcium silicate hydrate (C-S-H), which enhances the strength of the hardened concrete, and improves its durability and appearance. Coarse and ground slag particles are shown in **Figure 5.**

Material	Ib./yd^3	Kg/m^3	Percent (by weight)
Portland cement	1200	712	28.5
Fine sand	1720	1.020	40.8
Silica fume	390	231	9.3
Quartz flour	355	211	8.4
HRWR	51.8	30.7	1.2
Accelerator	50.5	30.0	1.2
Steel fibers	263	156	6.2
Water	184	109	4.4

Table 3.
Typical composition of Ductal [11].

Figure 5.
(A) Granular slag particles and (B) ground slag particles.

4. Mix proportioning, mixing, and curing procedures

Mix proportioning of concrete mixes containing one or more SCMs targets the optimization of mix packing order to minimize voids and increase strength, and durability without altering mix slump and/or flowing ability. Different proprietary UHPC mixes are available in the construction market. Examples of proprietary mixes are BSI "Beton Special Industrial," developed by Eiffage, Cemtec developed by LCPC, and Ductal developed by Lafarge. Different non-proprietary, ultra-high performance and high-performance mixes are developed through research programs across the globe.

The primary difference between UHPC and HPC mixes is the reduced strength of HPC (UHPC mix compressive strength exceeds 21.7 ksi [12, 13], and contains a higher percentage of SCMs, and incorporates random steel fibers to increase concrete ductility and post-cracking stiffness).

Supplementary cementitious materials and different steel fibers increase the overall density of concrete mixes. Optimized mixing procedures for concrete with high percentages of SCMs are as follows [14]:

I. Preblended the concrete mix granular ingredients for 2–3 minutes. The granular ingredients include cement, SCMs, fine sand, and coarse aggregates (if present).

II. Mixing water is added to the preblended powder. Mixing water may include 50% of the high-range water reducers (HRWR) included in the mix design. Wet mixing should continue for 8–10 minutes.

III. The remaining HRWR is added to the mix during the wet mixing process.

IV. Steel fibers included in UHPC are added before wet mixing is ended.

Due to the high packing order of granular materials and the possible inclusion of random steel fibers, high-energy paddle mixers are required to produce mixes with sufficient rheology. Examples of high-energy mixers are shown in **Figure 6**.

Concrete mixes with incorporated SCMs can be cured using regular moisture curing techniques. When very high early strength is required (as in precast/prestressed concrete industry), thermal curing could be applied. Thermal curing should be gradually applied to avoid developing hair cracks within hardened concrete [14].

Figure 6.
High-energy paddle mixers—Used for UHPC and HPC mixing.

4.1 SCMs in UHPC concrete

Silica fume and quartz flour are used in proprietary UHPC mixes to increase mix strength and durability. Current proprietary mixes, commercially available in the US and EU markets, have a high content of SCMs. In addition to their strength and durability advantages, SCMs partially reduce the cement consumption and hence reduce the carbon footprint [15–17]. Mix design and SCM content in major proprietary mixes are shown in **Table 4**.

The compressive strength of proprietary UHPC mixes ranges from 160 to 200 MPa. The strength is significantly increased due to the incorporation of SCMs and steel fibers. Steel fibers result in improved tensile capacity and higher values for the modulus of elasticity (MOE) and modulus of rupture (MOR). Detailed mechanical properties of proprietary UHPC mixes are shown in **Table 5**.

The high strength of girders produced using UHPC and HPC concrete mixes enables design engineers to fabricate shallow girders with a very high span-to-depth ratio. A comparison of sections with similar capacity produced by different materials including proprietary UHPC mixes is shown in **Figure 7** [29].

	Ductal	**Cemtec**	**Cor-Tuf**	**CRC**
Portland cement	712	1050	790	861
Silica fume	231	268	308	215
Quartz flour	211	—	216	215
Total cementitious materials	1154	1318	1314	1291
Percentage of SCMs to C. materials	38%	20%	40%	33%
Sand	1020	514	765	792
Water	109	188	166	220
HRWR	30.7	44	14	9.45
Accelerator	30	—	—	—
Fibers	156	180	166	218
W/CM	0.21	0.17	0.14	0.18

Table 4.
Proprietary UHPC mix designs in global construction markets [18–20].

	Value	**ASTM Standard [22–28]**
Design compressive strength	Greater than 150 MPa	ASTM C39/C39 M
Flexural strength	Greater than 20 MPa	ASTM C78/C78M-18
First cracking strength	Greater than 4 MPa	ASTM C1018–97
Creep coefficient	0.2	ASTM C512/512 M-15
Linear expansion coefficient	12×10^{-6}	ASTM C531–18
Elastic modulus	45 GPa	ASTM C469/C469M-14
Spread (flowing ability)	55 to 75 cm	ASTM C1611M-18

Table 5.
Average mechanical properties of proprietary UHPC mixes [21].

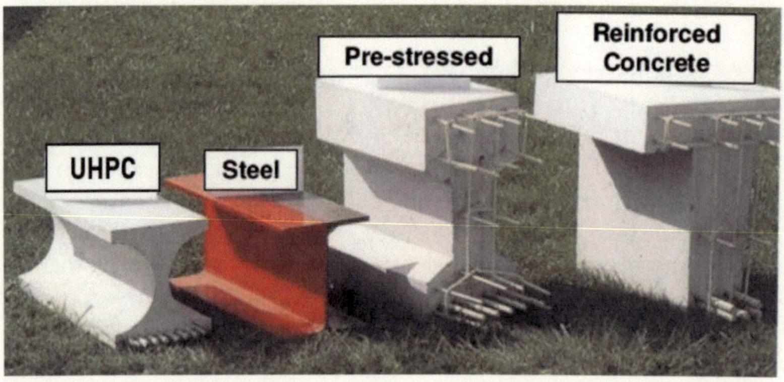

Figure 7.
Different girder sizes providing similar flexure capacity [29].

4.2 SCMs in HPC mixes

HPC mixes incorporates different SCMs with variable ratios according to the mix design purpose. Silica fume is used to increase the binder content and improve mix mechanical properties, whereas fly ash is primarily used to increase flowing ability. HPC mixes are produced using similar batching, mixing, and curing procedures as compared to UHPC. Steel fibers are eliminated due to their high cost, while chemicals for increased flowing ability are used to maintain high flowing ability using a low water-to-powder ratio. Different HPC mixes are shown in **Table 6**.

The aforementioned non-proprietary mixes had an average 24-h compressive strength of 80 MPa, and a final 28-day compressive strength of 110 MPa. Detailed compressive strength testing results are shown in **Figure 8**.

Current codes and standard specifications provide equations to estimate concrete mechanical properties as a function of its compressive strength. The American Concrete Institute (ACI) 318 calculates the modulus of elasticity of concrete (MOE) according to the following equation:

$$E_c = 0.043\, w_c^{1.5} \sqrt{f'_c} \qquad \text{(MPa)} \tag{3}$$

	Mix #1	Mix #2	Mix #3	Mix #4	Mix #5
Portland cement	630	625	630	670	630
Silica fume	90	80	90	145	90
Class C fly ash	180	80	180	145	180
Total cementitious materials	900	785	900	960	900
Percentage of SCMs to C. Materials	30%	20%	30%	32%	30%
Sand	1350	1450	950	1350	950
Water	135	155	145	145	140
HRWR	37	21	37	43	43
Fibers	—	—	—	—	—
W/CM	0.18	0.22	0.20	0.19	0.19

Table 6.
Non-proprietary HPC mixes using local materials in the US market.

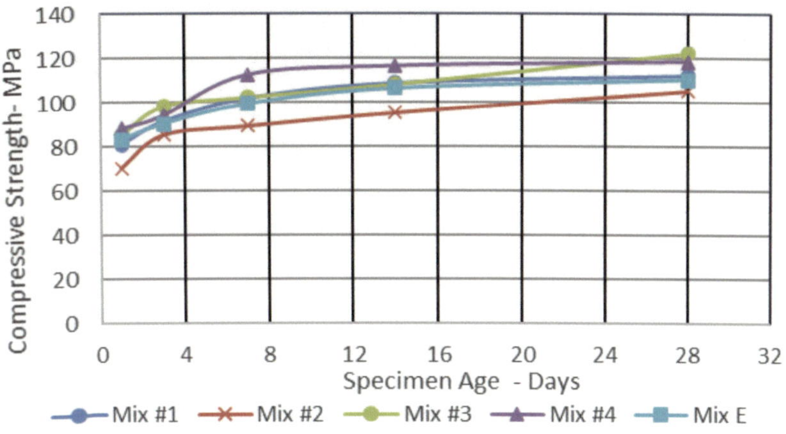

Figure 8.
Compressive strength test results for non-proprietary HPC mixes.

Similarly, ACI 318 and AASHTO LRFD specifications use the following equation to estimate the modulus of rupture (MOR) of concrete as

$$f_r = 0.62\sqrt{f'_c} \qquad \text{(MPa)} \tag{4}$$

Eqs. (1) and (2) show that the MOE and MOR of concrete, denoted as E_c and f_r respectively, and are correlated with the concrete compressive strength, denoted as f'_c. Thus, the increase in concrete strength associated with the use of SCMs in mix development results in a significant increase in concrete MOE and MOR.

5. Applications of concrete incorporating SCMs

Concrete mixes, with different SCMs, are currently used in different applications that requires high early strength, superior mechanical properties, and increased durability. The following represents main applications in residential and heavy construction applications:

5.1 High-rise residential construction

HPC, including SCMs, mixes are currently used in the construction of structural members in high-rise buildings. The use of HPC in high-rise construction started in the 1970s in the metropolitan areas within the United States including New York, Los Angeles, and Chicago. Recently HPC mixes with the high flowing ability (SCC characteristics) are used in pumping concrete floors in the City of Jeddah's Kingdom Tower and the Iconic Tower in Egypt's New Administrative Capital (shown in **Figure 9**).

5.2 Precast/prestressed bridge construction

Prefabricated prestressed girders with spans up to 60 meters are increasingly used in bridge construction, including accelerated bridge construction projects [30]. In order to increase the productivity of prestressing facilities, very high early strength is required before strands are released. Some prestressing facilities require

Figure 9.
Iconic tower—Egypt's new administrative capital.

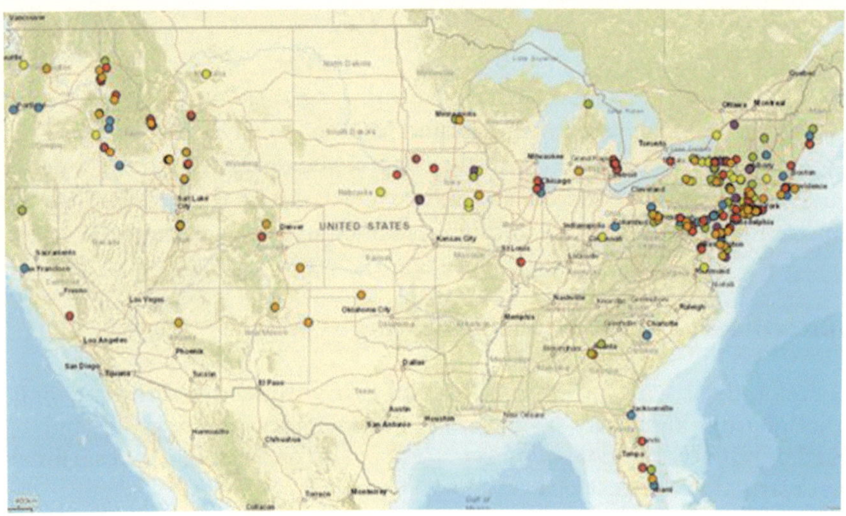

Figure 10.
UHPC bridge construction projects in the United States [32].

24-h compressive strength in excess of 70 MPa to release larger strands of 15- and 18-mm. diameter. Thus, SCMs, mainly silica fume, are incorporated in the concrete mix to ensure high early strength due to increased binder content and avoid girder cracking upon strand release [31]. UHPC is increasingly used in the construction industry, especially in heavy construction applications, with emphasis on long-span precast/prestressed girder bridges construction. In early 2000s, the first UHPC bridge was built in the United States using Ductal concrete (with 38% SCMs content) and steel fibers incorporated in the mix. The first UHPC bridge—known as Mars Hill Bridge—was constructed in the state of Iowa using FHWA funding. The successful construction of the Mars Hill bridge and the attained advantages resulted in the construction of a large number of UHPC bridges, mainly on the east coast of the United States, are shown in **Figure 10**.

Figure 11.
UHPC architectural pours.

5.3 Architecture applications

SCMs are used in producing mixes with very high flowing ability and early high strength for architecture construction. The high flowing ability allows for the poor of complicated shapes and limited thicknesses. Architectural UHPC pours are shown in **Figure 11**.

6. Challenges of SCMs in concrete industry

The use of SCMs in concrete mix development is faced by several challenges that limit their use in construction projects. Major impediments to the widespread of SCMs incorporation in concrete mix design include the following:

I. Material costs for specific SCMs are significantly higher than conventional concrete mix constituents. The average cost of micro-silica is $900 per ton in the United States market. In addition, micro-silica availability production is limited to few states, which may result in increased shipping cost and extended lead time.

II. Special mixers and mixing regimen are required, which necessitate modernization of batch plants and precast facilities to include SCMs in concrete mixing including changes in storage, batching techniques, and the type/size of mixers used.

III. Impact on fresh concrete mix properties due to the different sizes of SCMs and different chemical composition as compared to conventional mix constituents as cement and sand. Additional SCMs result in lowering average particle size for the mix and increase the density, while significantly reducing voids and permeability. These chemical and volumetric changes may require additional provisions to avoid the development of concrete lumps within the mixer. Provisions include the use of high-range water reducers (superplasticizers) and extend mixing time, and use special types of high-energy mixers.

IV. Health and safety restrictions should be followed when SCMs are used to avoid inhaling fine particles. Continuous exposure to certain types of SCMs as micro-silica might result in serious injuries or illnesses as silicosis. Thus, special PPEs are required when SCMs are used in mix development.

7. Conclusion

Supplementary cementitious materials (SCMs) are increasingly used in the construction industry to develop mixes with superior mechanical properties and improved long-term performance (durability). Different types of SCMs are available in the global market including micro-silica, class C fly ash, class F fly ash, quartz flour, and blast furnace slag. SCMs are incorporated in mix designs in partial replacement of Portland cement, with percentages ranging from 10% up to 40%. SCMs are used in proprietary UHPC mixes as Ductal, Cemtec, CRC, and Cor-Tuf, with percentages ranging from 20–40%. In non-proprietary HPC mixes, SCMs would have similar percentages; however, final strength is reduced due to the absence of steel fibers. SCMs improves the concrete properties through three techniques: 1) increase the amount of binder resulting from the hydration process, which increases the concrete final strength, 2) react with calcium hydroxide, which stops the efflorescence phenomenon, and 3) increase the packing order (density) of the cement matrix, which reduces concrete permeability, and its susceptibility to environmental attacks, de-icing salt effect, and protect reinforcing steel against corrosion. The reduction of cement consumption, being partially replaced by SCMs, reduces the carbon footprint of the construction industry.

The advantages of SCMs are partially offset with challenges including the scarcity of specific types of SCMs in some parts of the world. Specific types of SCMs are expensive compared to cement. Also, a special mixing regimen is required to avoid losing mix-flowing ability. Currently, SCMs are successfully used in concrete mix designs for high-rise residential construction, precast/prestressed girder bridge construction, and in architecture applications. The continuous research in concrete mix development using alternative SCMs will result in increased market share in global construction markets.

Acknowledgements

The author would like to acknowledge Professors Maher Tadros and George Morcous, University of Nebraska-Lincoln, for their invaluable technical advice. The author would like to acknowledge Professor Micah Hale, University of Arkansas, for his support. Finally, the author would like to acknowledge the generous material donation of Holcim Cement, BASF Master Builders, Chryso, Euclid Chemicals, and Martin Marietta for chemicals and aggregates supplies.

Conflict of interest

The authors declare no conflict of interest.

Author details

Amin Akhnoukh
East Carolina University, Greenville, NC, United States of America

*Address all correspondence to: akhnoukha17@ecu.edu

References

[1] Akhnoukh AK, Soares R. Reactive powder concrete application in the construction industry in the United States. In: Proceedings of the 10th Conference of Construction in the 21st Century (CITC 10), Sri Lanka; 2008

[2] Perry VH. What really is ultra-high-performance concrete? – Towards a global definition. In: Proceedings of the 2nd International Conference on Ultra-High-Performance Concrete Material & Structures (UHPFRC), China; 2018

[3] Akhnoukh A, Buckhalter C. Ultra-high-performance concrete: Constituents, mechanical properties, applications, and current challenges. Case Studies in Construction Materials. 2021;**15**:e00559. DOI: 10.1016/j.cscm.2021.e00559

[4] Akhnoukh AK, Kamel LZ, Barsoum MM. Alkali-silica reaction mitigation and prevention measures for Arkansas local aggregates. World Academy of Science, Engineering, and Technology International Journal of Civil and Environmental Engineering. 2016;**10**(2). DOI: 10.5281/zenodo.1338860

[5] ACI Committee 116. Cement and Concrete Terminology (ACI 116R-00) (Reapproved 2005). Farmington Hill, Michigan: American Concrete Institute; 2005

[6] Akhnoukh AK, Massarra C, Meadati P. Self-X concrete applications in smart cities. In: Proceedings of the Joint International Conference on Design and Construction of Smart Cities. 2019

[7] Hassan M, Dylla H, Mohammed L, Rupnow. Evaluation of the durability of titanium dioxide photocatalyst coating for concrete pavement. Construction and Building Materials Journal. 2010;**24**(9):1456-1461

[8] Khayat KH. Workability, testing, and performance of self-consolidating concrete. ACI Materials Journal. 1999; **96**(3)

[9] https://gcpat.com/en/solutions/products/tb-0709-force-10000-d-microsilica-and-its-uses-concrete [Retrieved: 08 June 2021]

[10] Akhnoukh AK. Development of high performance precast/prestressed bridge girders [dissertation]. Nebraska, USA: University of Nebraska-Lincoln; 2008

[11] FHWA. Ultra-high-performance concrete: A state-of-the-art report for the bridge community. Federal Highway Administration. Publication No. FHWA-HRT-13-060. 2013

[12] Toutlemonde F, Resplendino J. Designing and building with UHPFRC. DOI:10.1002/9781118557839

[13] JSC. Recommendations for design and construction of high-performance fiber reinforced cement composites with multiple fine cracks (HPFRCC). Japan Society of Civil Engineers. 2008

[14] Akhnoukh AK, Elia H. Developing high performance concrete for precast/prestressed concrete industry. Case Studies in Construction Materials. 2019; **11**:e00290. DOI: 10.1016/j.cscm.2019.e00290

[15] Elia H, Ghosh A, Akhnoukh AK, Nima ZA. Using nano- and micro-titanium dioxide (TiO_2) in concrete to reduce air pollution. Journal of Nanomedicine and Nanotechnology. 2018;**9**(3). DOI: 10.4172/2157-7439.1000505

[16] Akhnoukh AK. Overview of nanotechnology applications in construction industry in the United States. 2013;**5**(2)

[17] Akhnoukh AK. Implementation of nanotechnology in improving the environmental compliance of construction projects in the United States. 2018;**36**(3):357-361. DOI: 10.1080/02726351.2016.1256359

[18] Rossi P, Arca A, Parant E, Fakhri P. Bending and compressive behaviors of a new cement composite. Cement and Concrete Research Journal. 2005;**35**: 27-33. DOI: 10.1016/j.cemconres.2004.05.043

[19] Aarup B. CRC – A special fibre reinforced high performance concrete. In: Proceedings of the First International RILEM Symposium on Advances through Science and Engineering. Publication Pro 048. 2004

[20] Howard LL, Carey A, Burcham M, Scott DA, Shannon JD, Moser RD, Horstemeyer MF. Mechanical behavior of cortuf ultra-high-performance concrete considering aggregate and paste effect. US Army Corps of Engineers Report # 18-31, 2018. DOI: 10.21079/11681/29642

[21] Graybeal B. Ultra-high-performance concrete: A state-of-the-art report for the bridge community. Federal Highway Administration Report FHWA-HRT-13-060. 2013

[22] ASTM C39/C39M-01. Standard test method for compressive strength of cylindrical concrete specimens. ASTM International; 2001

[23] ASTM C78/C78M-21. Standard test method for flexural strength of concrete (using simple beam with three-point loading). ASTM International; 2021

[24] ASTM C1018 – 97. Standard test method for flexural toughness and first-crack strength of fiber-reinforced concrete (using beam with third-point loading). ASTM International; 1997

[25] ASTM C512/512M-15. Standard test method for creep of concrete in compression. ASTM International; 2015

[26] ASTM C531-18. Standard test method for linear shrinkage and coefficient of thermal expansion of chemical-resistant mortars, grouts, monolithic surfacings, and polymer concrete. ASTM International; 2018

[27] ASTM C469/C469M-14. Standard test method for static modulus of elasticity and poisson's ratio of concrete in compression. ASTM International; 2014

[28] ASTM C1611M-18. Standard test method for slump flow of self-consolidating concrete. ASTM International; 2018

[29] https://www.ductal.com/en/engineering/why-use-ductal [Retrieved on 06/12/2021]

[30] Akhnoukh AK. Accelerated bridge construction projects using high performance concrete. Case Studies in Construction Materials. 2019;**12**. DOI: 10.1016/j.cscm.2019.e00313

[31] Akhnoukh AK. The effect of confinement on transfer and development length of 0.7-inch prestressing strands. In: Proceedings of the 2010 Concrete Bridge Conference: Achieving Safe, Smart & Sustainable Bridges. Arizona, USA. 2010

[32] https://usdot.maps.arcgis.com/apps/webappviewer/index.html?id=41929767ce164eba934d70883d775582 [Retrieved on 06/12/2021]

Chapter 9

Application of a Granular Model to Identify the Particle Size of the Granular Mixtures of Concrete Based on Dune Sands

Mhammed Abdeldjalil

Abstract

The control of the determination of concrete depends on the basic properties of the desired concrete and thanks to the type of granular mixture of concrete. We arrive at the required concrete quality. And in this study, we can identify the granular distribution class of concrete using the fractal model. In particular, the granular distribution can be determined by the fractal dimension, either for each granular component separately, or for the dry granular mixture of the concrete. The fractional dimension is obtained by transforming the particle size curve to a fractal line. In this study, we used some experimental results obtained from projects already carried out in arid regions. Knowing that we have applied parameters such as granular extent and fractional dimension to the study of these existing projects, we can define a dry mix of concrete through the granular distribution. Therefore, we used the program that we proposed previously of transforming the grain size curves to a fractal line which was obtained for each grain mixture with a very acceptable correlation.

Keywords: Concrete, Aggregate, Fractal Dimension, Granularity, Identification

1. Introduction

Civil engineering practitioners such as researchers, engineers, technicians and those interested in it have always favored the best concrete formulas based on classical or modern methods. The methods used were either experimental (laboratory tests), or empirical or semi-empirical method or analytical methods. The current results of concrete production always indicate that they are oriented towards modern methods which are mainly based on numerical modeling [1–10].

Scientific research in the field of civil engineering, as in many other applied sciences and technologies, is a major use of its implementation according to modern tools, in particular calculation tools [1, 3]. The main objective of this study is to discover the best methods that help in the economic aspect, in particular the building materials when testing, in addition to saving time and effort lost in their realization without taking them into account. Thus, it was more interesting to use technology (electronics and computer) with analytical methods and concrete models to achieve experimental methods.

IntechOpen

On the other hand, our main goal is to take advantage of recent studies [5–10] that rely on granular distribution to quantify aggregate dosages. According to the parameters of the fractal dimension FD [1–4] and the granular extent D/d [2].

On the other hand, we emphasize the importance of extending the determination of the granular distribution by using the fractal distribution as a new model to determine the granular mix class of concrete. Our experimental results, which we adopted in this study, allow us to determine a numerical value as one of the physical properties of the aggregate, which is the fractal dimension (FD). It facilitates the formulation of concrete by precisely specifying the components of the aggregate. Our objective in this study is to create a large database which helps us to save time and materials in experimental studies within the framework of concrete formulation methods to determine dosages of granular materials, which is certainly useful in the field of civil engineering. Initially, we rely on the study of a component to facilitate the process and start from the simplest operations. We took, for example, the study of the effect of the sand component [4]. It should be noted that this work is mainly based on the data, which takes the aggregate as the basic component in the production of concrete so that as it is known, 80% of the concrete is composed of aggregate, in order to obtain a good granular distribution (continuous granulometry). We emphasize that the appropriate selection of aggregates according to the desired concrete requirements allows us to achieve one of the most important characteristics that distinguish concrete, which is the compressive strength of concrete [3, 4]. In addition, the new definition of granularities by fractal dimension helps in choosing the classes of aggregate to be used in concrete. We confirm that one of the main objectives of this study is to highlight the optimal importance of the fractal dimension parameter and its results, which brings us to the possibility of knowing how to determine the reference particle size curve of granular mixtures at using the fractal model for granular concrete mixtures.

2. Conventional concrete methods

To obtain a concrete having the desired properties according to climatic and other requirements, and to use local materials for economic reasons and in order to know the proper method of concrete formulation, we have seen that it is necessary to mention some of the conventional methods widely used in the formulation of concrete.

Methods were adopted for the formulation of concrete whose first principles of physical relations emerged at the end of the 19th century, and these methods of formulation have varied depending on the materials available and our need for the required concrete quality.

2.1 Strength formulas

2.1.1 Formulas of Feret

René Féret [11] in 1892 was one of the first to research the law governing the prediction of the compressive strength of concrete f_c (1).

Its formula based on the strength of the cement (the true class), the nature of the aggregate, the cement/water dosage ratio and taking into account the volume of voids. But does not take into account neither the shape of the aggregate nor the granular distribution, nor the resistance to fragmentation of the aggregate. The latter is formulated using the following expression:

In 1892, Féret [11] to whom the first researches are attributed, worked on a principle of the mechanical resistance of concrete f_c (1). K_{feret} coefficient of the resistance of the cement and the type of the aggregate. c, e are dosages of cement and water. v is the volume of the area. But it does not directly take into account neither the shape nor the type of gravel used, as well as the granular distribution of the granular concrete mixture.

$$f_c = K_f . f_{mc} \left(\frac{V_c}{V_c + V_w + V_a} \right)^2 \tag{1}$$

- f_c: Strength of the concrete at the maturity considered
- K_f: Model constantly (index f for Féret)
- f_{cm}: Normal strength of cement
- V_c Absolute volume occupied by the cement
- V_w: Volume occupied by water
- V_c: Air volume

2.1.2 Methods of fuller

Fuller and Thomson [12] in 1907 established their method based on the maximum compactness of the continuous granular mixture, and it depends mainly on the porosity of the granular mixture (2) and the granular expansion. However, it does not directly take into account the shape of the grains, nor the resistance to friability of the aggregate used, and its relation is written as soot:

$$P_{F\,T} = 100 \sqrt[5]{\left(\frac{d}{D} \right)} \tag{2}$$

- P_{FT}: Porosity of the granular mixture
- d/D: The granular extension.

2.1.3 Methods of Abrams

Abrams [13] in 1918, Regardless of the European school, he empirically proposed an exponential equation to predict the compressive strength of concrete, still used in North America, which has two adjustable parameters [Popovics, 1995].

The cement/water ratio, and involves through a coefficient (improved K_Féret) which indirectly presents nature and shape of the aggregates.

We note the absence of a direct representation of the resistance to fragmentation of the aggregate and of the granular distribution, as is the case in the rest of the previous methods, its formula (3) is written:

$$f_c = K_{Abrams} \left(\frac{1}{7.5^{(1.5(W/C))}} \right) \tag{3}$$

- W/C: Ratio of cement and water,
- K_{Abrams}: Nature and the form of the aggregates,
- f_c: Mechanical strength of concrete

2.1.4 Methods of Bolomey

Bolomey [14] in 1925 is based on a formula (4), (improved iron) to determine the dosages of cement and water. This formula for predicting the mechanical compressive strength of concrete, which depends on the shape of the aggregates as well as the consistency of the concrete, and the dosages of cement and water, and the volume of voids. But does not take the representation of resistance to aggregate fragmentation.

This formula, like that of Féret, is the product of three terms which share, in order of factors, the influence of aggregates, cement and concrete formulation. The difference, compared to the relation of Féret, relates exclusively to the third term, parabolic in Féret, linear in Bolomey. It has been shown that the Bolomey relation is a good approximation of that of Féret for the values of the E/C ratio between 0.40 and 0.70; within this range, the error is less than or equal to 3%.

$$f_c = K_{Bolomey}\left(\frac{C}{W+V} - 0.5\right) \quad (4)$$

- f_c: The mechanical resistance of concrete,
- K_B: Depends on the shape of the aggregates and the consistency of the concrete
- C, W: The dosages of cement and water,
- V: The volume of the area.

2.1.5 Methods of Caquot

The scientist Caquot [15] circulated his research during the year 1937, through which he sought to find the optimal aggregate distribution in which the porosity of the aggregate mixture is minimal, according to the basic hypothesis of compatibility between two aggregates classes without influence due to the presence of another aggregates class.

This basic idea was taken up by F. de Larrard [3], who had previously embarked on a vast process of developing other concrete formulation programs.

The relation is determined empirically by assuming that the volume of the voids depends on the dimensions of the small grains, then on the addition of grains, then on a constant determined empirically according to the relation of Caquot (5).

$$V = V_0\sqrt[5]{\left(\frac{d}{D}\right)} \quad (5)$$

- V: The volume of voids,
- d, D: The dimensions of small grain, more grain.
- V_0: Constant defined experimentally

2.1.6 Methods of Faury

We find in the work of Faury [16] and Joisel [17] that they made modifications to the work of Caquot in 1942 and 1952, and Faury extends to the granular range up to 6.5 μm, incorporating the cement as a granular material and taking into account the effect of the wall. And Joisel gave a reference straight line (at a complex scale) taking into account the cement, water, voids, granulometry and the compactness of the granular classes. Here, we note an indirect representation of the granular distribution with the mechanical resistance of the aggregates [18].

The optimum grain size of a concrete is a mixture (in a certain proportion) of two kinds of grains of the aggregate.

The reference curve to be followed consists of two straight sections.

The first AB gives the granulometry of fine grains. The second straight line is that of coarse grains. The y coordinate of, called the break point, indicates the percentage by volume of the grains. Its value is given by the experimental formula (6).

$$Y = A + 17.\sqrt[5]{\left(\frac{d}{D}\right)} + \frac{B}{\frac{R}{D} - 0.75} \tag{6}$$

- D: maximum aggregate size in mm
- **R**: average radius of the formwork in mm
- A: coefficient taking into account the shape of the aggregates and the consistency of the concrete.
- B: depends on the tightening. It varies from 1 for a powerful vibration to 1.5 for an average tightening.

2.1.7 Methods of Dreux-Gorisse

He method of Dreux and Gorisse [19] is based on the optimal granularity which is still current for the design of the concrete formulation. This is an empirical approach according to an OAB granular reference curve (segments of two lines in a semi-logarithmic plot). Contrary to the moment, the cement is not part of the reference curve of the mixture, since its mass is determined separately. It is a method which takes into account a large number of parameters [18]. But it does not take into consideration the direct representation of the granular distribution of the aggregate, and indicates what the true class of cement represents, and the dosage of cement and water, type, shape, quality and dimensions. of aggregates, the smoothness, consistency and strength of concrete.

This method is fundamentally empirical in nature, unlike the Faury method which predates it [Faury, 1942] and which is based on Caquot's theory of the granular optimum [Caquot, 1937]. Dreux carried out a large survey to collect data on satisfactory concretes [de Larrard, 2000]. On the basis of a statistical analysis of this large number of concretes and by combining the granular curves obtained, they were able to base an empirical approach to determine a reference granular curve.

It is also very easy to use since it only requires knowing the grain size curves of the aggregates used.

A test batch is necessary to be carried out in the laboratory in order to make any usage corrections.

"B" (on the ordinate 100%) corresponds to the dimension D of the largest aggregate.

"O" (at ordinate 0%) corresponds to the dimension d of the smallest aggregate.

The break "A" has the following coordinates:

- on the abscissa (from the dimension D of the sieve) on the ordinate.

Si: $D \leq 20$ mm; the abscissa is D / 2.

If: $D \geq 20$ mm; the abscissa is located in the middle of the "gravel segment" limited by the modulus 38 (5 mm) and the modulus corresponding to D.

in ordinates (7)

$$Y = 50 - \sqrt{D} + K + K_s + K_p \tag{7}$$

- *K*: corrector which depends on the cement dosage, clamping efficiency, the shape of the rolled or crushed aggregates

- K_s additional correction according to the fineness modulus of the sand (case of coarse sands) by adding the value $K_s = 6.M_F - 15$

- K_p: the coefficient depends on pumped concrete or not.

2.1.8 Baron and Lesage

The method of Baron and Lesage [20] is based on a technique proposed in 1976 to improve the granular skeleton according to the principle of relating the minimum flow time specified by the LCL Maniabilimeter according to Standard 18–452 [21] with the quantity optimal granularity for constant cement and water ratios.

The principle is to measure the time taken for a concrete sample to flow under vibration to a certain mark. The optimum proportions of aggregate are assumed to give the minimum flow time, for a given amount of cement and water. Once the granular proportions have been identified, the water and cement dosages are adjusted experimentally, so that the mixture has the desired workability and resistance. It is assumed, in this method, that the optimum proportions of aggregates do not depend on the quantity of cement.

2.2 Other models

Baron and Olivier have developed a formulation method derived from the Dreux-Gorisse method with modifications to adapt to existing concrete in 1996. This method makes it possible to integrate a large number of the parameters listed, taking into account the additives and incorporating mineral additives.

Baron and Olivier have developed a concrete formulation method derived from the Dreux-Gorisse method with modifications to adapt to existing concrete in 1996. This method makes it possible to integrate a large number of listed parameters, taking into account the additives and incorporating mineral additions.

The method is due to Mr. BARON, from the experimental studies he carried out in the years 1970–1980 and which were subsequently optimized by using the directives of the NF P 18–305 standard, replaced by the European Standard. EN 206–1. This method is developed in a book co-written by Messrs. BARON and OLLIVIER "BETONS, Bases and data for their formulations" published by Eyrolles.

The experiments confirmed the work carried out by BOLOMEY and FAURY, certain formulas of which were adopted with regard to the binder and water dosages.

For the dosage in aggregates, it is, in part, the work of DREUX that was retained. The whole is completed by experimental results which make the method as affordable as the DREUX method without having the drawbacks of its limitation to the only common concretes.

The problem of the optimal dosage of concrete is not unique: there are actually two problems that can be addressed independently of each other:

- Binding paste

We start the formulation from 2 main assumptions which are the target resistance and the optimal effective water quantity.

The target resistance R_c is obtained from the calculation of the concrete or R_{c28} required by the work to be constructed. Taking into account the true resistance class of the cement and the nature of the aggregates, the BOLOMEY formula is used to define the W/C ratio. The optimum effective water is defined in a simple and provisional way according to the target consistency of the concrete by a table created by BARON taking into account a certain number of corrections relating to the dimension D of the gravel used (dimension of the smallest sieve which leaves pass all the components of the concrete) and at the temperature of the concrete pour.

From these values, we can therefore determine the cement dosage. Corrections are made from a trial mess.

- The granular skeleton

The granular skeleton retained by BARON is very close to that obtained by DREUX, however with a simpler approach and definition. This method was chosen not for its scientific basis, but because its results have been satisfactory over the past 25 years.

There are other innovative methods of concrete formulation, which can use numerical models such as René LCPC or BétonLab and BétonLab Pro2 [3], and other methods are analytical.

The first theories dealing with the maximum pressure of granular mixtures [Féret 1892, Caquot 1937] do not explicitly take into account the interactions and grain sizes between them. Through the compact stacking model, De Larrard [3] incorporates new concepts such as clamping, wall effect and thinning effect.

3. New models (mathematical analysis)

3.1 Fractal analysis

The fractal model for determining the granular distribution is a conclusion drawn from fractal analysis and is a new model for the mathematical description of everything used and found in nature, in which its truncated shapes reveal patterns similar to increasingly precise scales. And irregular and recurring shapes can be described using mathematical models. The term "fractal" is a mathematical term coined by Benoit Mandelbrot [22] from the Latin root fraction. It was originally used as an adjective (fractal line) and today is a noun meaning broken or irregular.

Sebsadji and Chouicha in 2012 [5–10, 23] showed that Fractals can be defined as disordered systems that are self-similar independent of scale of observation. Their fundamental property is a non-integer dimension called fractal dimension, which can measure.

Its result builds on the findings of previous researches (Lecomte and Thomas, 1992; Chouicha, 2006), according to which ideal grading curves of concrete can be transformed into straight-lines power-law of the form given in Eq. (11).

3.2 Fractal dimension (FD)

Marmi [24], in 2019 he expressed the fractal dimension as a parameter exists in classical geometry, and is a line is a one-dimensional object a surface a two-dimensional object, a volume a dimensional object. We are therefore used to objects whose dimension (D) is an integer 1,2 or 3. But it is not specified, what would be the dimension of a series of points on a line, an irregular and plane curve, a surface full of convolutions. For this purpose, the term fractal dimension was introduced by B. Mandelbrot in 1975 the fractal dimension is therefore a number which measures the degree of irregularity or fragmentation of an object or which measures the roughness of a surface.

The fractal dimension is the fraction or an irrational number (; 1.23; etc.) or an integer.

This notion of fractal dimension applies to scale-invariant objects: there are parts which are similar to the object itself up to an expansion (enlargement).

When we change the observation scale of a scale invariant object, we keep the shapes.

The particle size distribution curves of the cumulative sieve percentages as a function of the grain dimensions can be transformed to a straight line representing cumulative numbers as a function of the grain dimensions.

We can do this by assuming that the shapes of the grains have the same oval shape, and this, if we adopt the same hypothesis proposed by Lecomte and Thomas [25] in his work, which first touched on the analysis fractal and through which he approached the application of the fractal dimension in the determination of three types of granular mixtures for high-performance concrete. And in 1992, he achieved his study results by applying fractal analysis to a granular mixture of concrete related to the definition of granular analysis of granular mixture of concrete, which consists of several granular types. These results indicate at the time that he adopted the hypothesis of the dimension of a spherical grain of aggregate of main and standard dimension G, and the relation (8) below summarizes the determination of the volume of the spherical grain.

We can estimate the mass of the grains, called **M** mass, as well as their true density **ρ** and the average size of the grains **V** for each initial class of grains, n being the partial number of grains refused on the opening screen **G** [25] as a relationship (9) below.

$$\boldsymbol{v} = \frac{\boldsymbol{\pi}}{\boldsymbol{6}}\boldsymbol{G}^3 \tag{8}$$

$$\boldsymbol{N} = \frac{\boldsymbol{M}}{\boldsymbol{\rho}\left(\frac{\pi}{6}\right)\boldsymbol{G}^3} \tag{9}$$

It is also possible to express the cumulative number Nc of aggregate grains whose dimension is greater than or equal to the size of the opening of the sieve G, and Relation No. (10) shows the determination of the cumulative number of grains of aggregate. Thus, Relation (11) allows us to express the number of grains of aggregate rejected in a sieve, in terms of the cumulative numbers of all the granular components.

$$\mathrm{N_c} = \sum_{i=1}^{n} \frac{Mi}{\rho\left(\frac{\pi}{6}\right)Gi^3} \tag{10}$$

$$N_c = \sum_{i=1}^{n} \frac{MiP}{\rho \frac{\pi}{6} Gi^3} N_{cumule} = \sum_{i=1}^{n} \frac{MiP}{\rho \frac{\pi}{6} Gi^3} \tag{11}$$

The granulometric analysis of cement is done by laser, "Laser granulometry" this technique is based on the diffraction of light and was proposed by Fraunhofer under the application of their theory of Fraunhofer ". We have **Table 1** below which shows by a sub-detail of the fractal analysis which will identify the particle size of an example of CPA cement by the fractal dimension (FD).

FD fractal dimension is, therefore, an approximation of the granular distribution curve. If this approximation is good over almost the entire grain size measurement field, the granular distribution line is said to be the fractal or quasi-fractal dimension. If the curve obviously tends towards a limit when the dimension of the seeds tends towards zero, then this curve is said to be semi-fractal.

We show without difficulty that any physical measurement on a granular structure, even purely fractal, results in a granular curve on a logarithmic scale (quasi-fractal) due to the smaller dimension of d mm, an empirically necessary procedure. It turns out that only successive zooms, and logarithmic scale transformations

Sieve size (mm)	Particle size analysis				Fractal analysis			
	Refusal mass (g)		% Cumulated		Density (g/cm³)	Grain volume (cm³)	Number of grains	
	Partial	Cumulated	Refusal	Passing			Partial	Cumulated
0.125	0.00	0.00	0	100	3,00	1.02E-06	0.00E+00	0.00E+00
0.1	2.40	2.40	1	99		5.23E-07	1.53E+06	1.53E+06
0.08	9.60	12.00	5	95		2.68E-07	1.19E+07	1.35E+07
0.063	12.00	24.00	10	90		1.31E-07	3.06E+07	4.40E+07
0.05	12.00	36.00	15	85		6.54E-08	6.11E+07	1.05E+08
0.04	12.00	48.00	20	80		3.35E-08	1.19E+08	2.25E+08
0.0315	24.00	72.00	30	70		1.64E-08	4.89E+08	7.14E+08
0.025	21.60	93.60	39	61		8.18E-09	8.81E+08	1.59E+09
0.02	14.40	108.00	45	55		4.19E-09	1.15E+09	2.74E+09
0.016	12.00	120.00	50	50		2.14E-09	1.87E+09	4.61E+09
0.0125	24.00	144.00	60	40		1.02E-09	7.83E+09	1.24E+10
0.01	12.00	156.00	65	35		5.23E-10	7.64E+09	2.01E+10
0.008	7.20	163.20	68	32		2.68E-10	8.96E+09	2.90E+10
0.0063	4.80	168.00	70	30		1.31E-10	1.22E+10	4.13E+10
0.005	4.80	172.80	72	28		6.54E-11	2.45E+10	6.57E+10
0.004	19.20	192.00	80	20		3.35E-11	1.91E+11	2.57E+11
0.00315	9.60	201.60	84	16		1.65E-11	1.96E+11	4.52E+11
0.0025	2.40	204.00	85	15		8.18E-12	9.78E+10	5.50E+11
0.002	2.40	206.40	86	14		4.19E-12	1.91E+11	7.41E+11
0.0016	9.60	216.00	90	10		2.14E-12	1.49E+12	2.23E+12
0.00125	2.40	218.40	91	9		1.02E-12	7.83E+11	3.02E+12
0.001	2.40	220.80	92	8		5.23E-13	1.53E+12	4.55E+12

Table 1.
Particle size analysis and fractal analysis of CPA.

(d, D), probably reveal (on the slope of the lower higher convergence line) the effective quasi-fractal drift of the studied process.

If this drift has several changes in the slope, then in some cases it will be referred to as "multi-fractal."

Figure 1 Presented the granular distributions of four types of cement identified by the fractal line, the cements are:

Portland cement compound class 42.5 MPa CPJ 42.5.

Cement sulphate resistant class 42.5 MPa CRS 42.5.

Artificial Portland cement class 52.5 MPa CPA 52.5.

Portland cement compound class 42.5 MPa type P6 CPJ P 42.5.

We present in **Figure 1** - the different fractal distribution with correlation coefficients of the fractal lines of the granular distributions, and the minimum correlation coefficient value is $R^2 = 0.96$. Appears in the granular distribution between three closely related types of cement, and another is different.

We followed the same method according to the results **Figure 2** of Lecomt [25] presented in the **Figure 2** which has ideally defined an example of the granular mixture containing a spread granular for a high-performance concrete, as well as all the granular classes of this concrete including the active mineral additions were used.

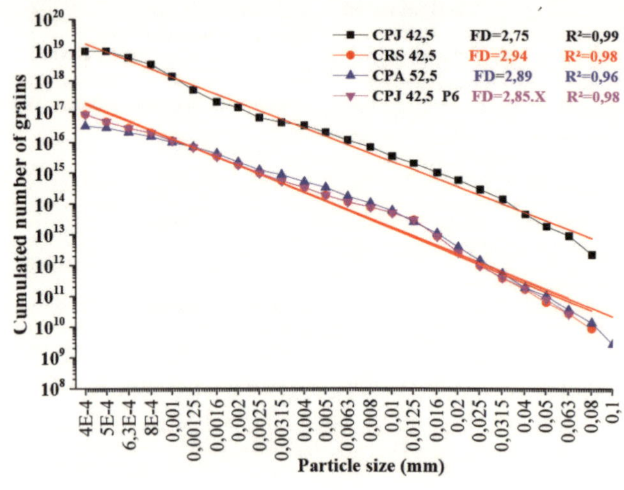

Figure 1.
Fractal lines of granular distributions of four types of cements alone.

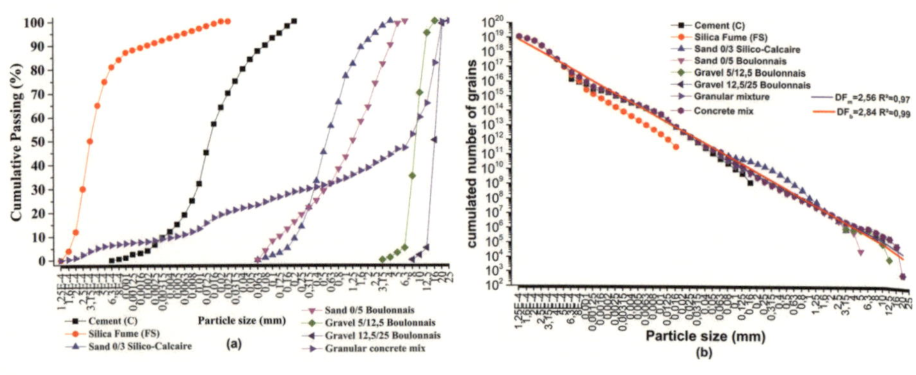

Figure 2.
Transformation from a particle size distribution to a fractal distribution for a concrete mixture and its components from the Lecompt [25]. a) Particle size curves, b) fractal line.

All concrete formulation methods, old or new, are based on particle size for determining the different dosages of the granular constituents. The results obtained by Lecompte [25] and Chouicha [2], show that these methods which use a granular distribution, indirectly use a fractal distribution.

In **Figure 3** of Chouicha [2], the particle size curves for different granular mixtures that he identified with a uniform particle size range, we can determine the particle distribution of the granular mixtures with a fractal dimension from FD = 0.5 to FD = 7, knowing that this field is for the granular mixtures in general, which is much larger than the field of the granular concrete mix, so it is outside the concrete field, because the fractal dimension FD does not exceed the value of 3 .

We applied one of the three high quality BHP concrete mixes that he adopted by Lecomt in his research (**Figure 2**), we clearly show through **Figure 4** the curves of the granular distribution of the component classes of the concrete, as well as its curve of the granular mixture, and this gives some similarities between Lecomt's work and what we got despite using different components in terms of density and type of aggregate ect, and what we got despite using different components in terms of density and origin of gravels, and this is due to our relying on the granule size assumption of the spherical-shaped relation (8) to obtain the fractal distribution of this granular mixture with its components as shown in **Figure 4 (a), (b)** which gives us the results of converting granular curves to fractal lines. Lines.

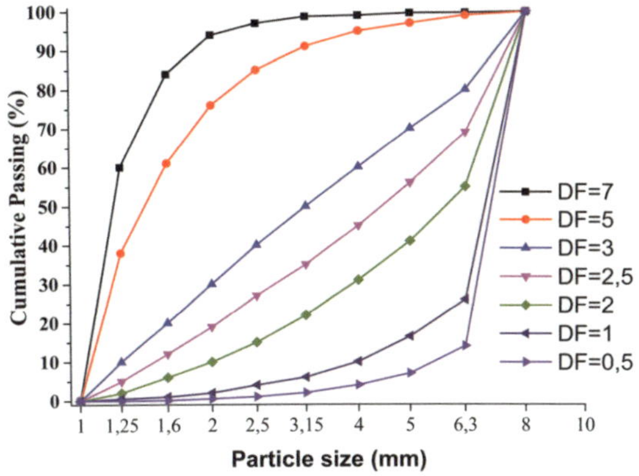

Figure 3.
Particle size curves of the different granular mixes identified by FD Chouicha [2].

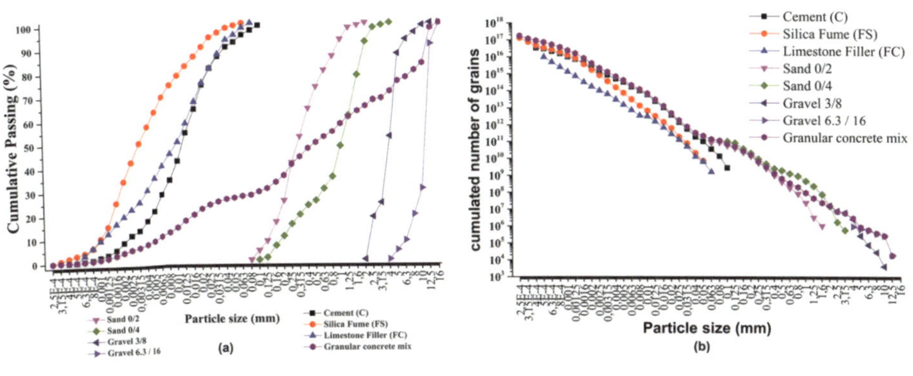

Figure 4.
Transformation from a particle size distribution to a fractal distribution for a concrete mixture and its components (example BHP) a) particle size curves, b) fractal line.

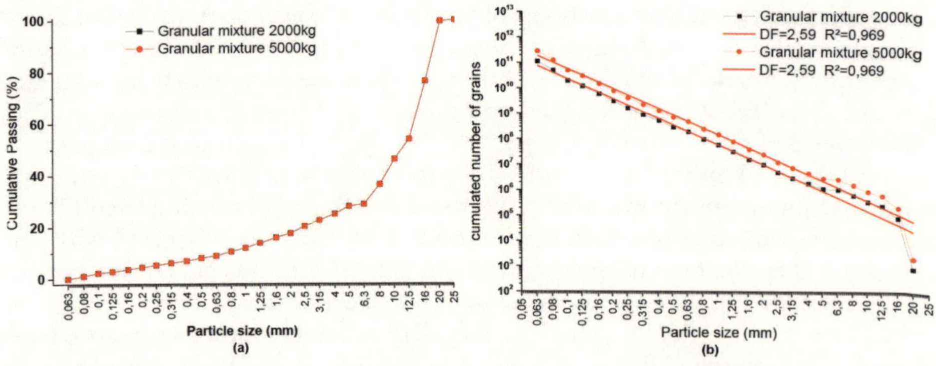

Figure 5.
Transformation of a particle size curve of a granular mixture 0,63/25 to a) fractal line particle size curves, b) fractal line.

The process of transforming the particle size curve into a fractal line has a direct relation to the granular variety of a granular class or of a granular mixture regardless of the mass taken for the granular variety. **Figure 5 (a)** shows that the grain size curve remains the same, regardless of the mass for the same grain class. **Figure 5 (b)** shows the transformation into a fractal line which gives us two lines of the fractal distribution, for each mass gives a fractal line for the same granular class, but with the same slope value, so it is the same value of the fractal dimension, and this is what he had confirmed by Chouicha [2] in his work in 2006.

4. The fractal line of sands/gravel and cement

First, on the one hand, we deal with the transformation of the granulometric curves of local materials concerning the granular classes of sand and then of gravel. On the other hand, we show the transformation of the particle curves for the cementitious materials of CPJ 42.5 and CRS 42.5.

The determination of the granular distribution is one of the important physical properties for the definition of aggregates and as is known in the context of concrete formulation methods, and it is important to control and determine the proportions of the appropriate aggregate components for concrete.

4.1 The fractal line of sands

Figure 6 (a) is an example showing the grain size curve of dune sand with a grain range of 0/5, **Figure 6 (b)** is its transformation into a fractal line, and this sand is one of the 10 sand classes of dunes shown in **Table 2** and its smallest, granular extent is 0.005/0.63 and it is a very fine dune sand. As for the granular extent of coarse dune sand is 0.005/5. Thus, a quarry sand with its granular extent is 0.063/63.

The identification results shown in **Table 2** were obtained by conducting laboratory experiments to determine the varieties of sand dunes at the Building Materials Laboratory of the University of Adrar in cooperation with the Regional Workshop Laboratory. On desert techniques (ARTS) [26]. The sand studied is sand from sites approved for the use of dune sands from different sites in the Saharan region in Adrar, Algeria.

The graphical fit of the fractal distribution is a linear fit by the equation $y = b + ax$. And constant "a ", it is the slope which represents the fractal dimension. **Figure 6** is

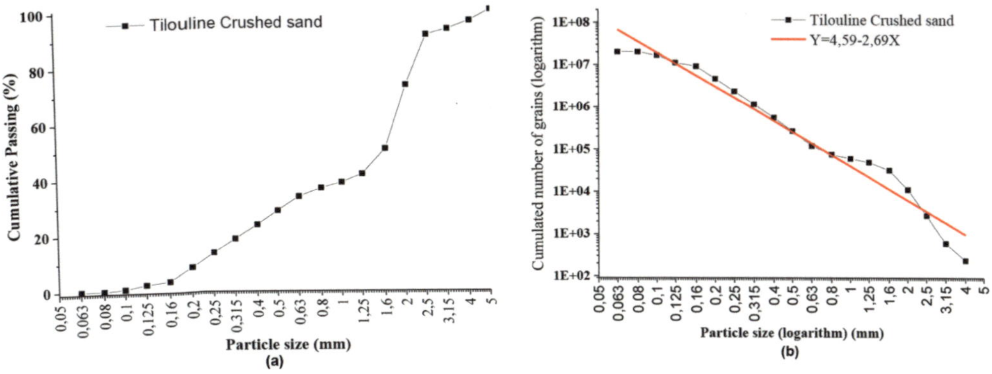

Figure 6.
Transformation of a grain size curve from a sand from Tilouline dune to a fractal line (example) [1]. a) Particle size curves, b) fractal line.

an example showing the methodology for determining fractal dimensions of all granular classes, including dune sand.

$$Y = 4,59–2,69\ (x) \tag{12}$$

In this case, the slope is 2.69 and the correlation coefficient is $R^2 = 0.96$. The slope obtained by cumulative number of grains according to grain dimensions is the fractal dimension DF, which determines the class of the distribution of grains of a granular class or of the granular mixture.

4.2 Physical properties of dune sands

Table 2 summarizes the conventional physical properties of dune sands in the Adrar region, namely granular extent, fineness modulus and sand equivalent. And the unconventional properties are the fractal dimension of the granular distribution, whose value varies from 1.14 to 4.20 with a correlation coefficient whose value varies from 0.82 to 0.98, which is a value close to one therefore is accepted. In addition to the total area of the grains of sand calculated by the fractal dimension.

Sand dune site	**ES**	**MF**	**STg (mm²)**	**d/D**	**DF**	**R^2**
Ouinna site n°01 - Adrar (SD)	82,43	1,78	8,58E+14	0,05/1,6	**4,20**	0,91
Tinerkouk - Timimoun (SD)	78,08	1,06	3,45E+14	0,05/2,5	**3,72**	0,89
Tilouline - Z. Kounta (SD)	82,17	1,42	2,34E+12	0,063/5	**2,69**	0,96
Ouinna site n°02 - Adrar (SD)	74,20	1,78	6,39E+11	0,063/0,8	**2,21**	0,89
Bordj Badji Mokhtar (SD)	73,18	1,64	3,38E+12	0,05/0,63	**2,86**	0,82
Tsabit - Adrar site n°01 (SD)	74,32	2,22	6,63E+13	0,063/2,5	**3,36**	0,93
Tsabit - Adrar site n°02 (SD)	74,32	2,22	1,50E+13	0,063/1,25	**3,88**	0,93
Tsabit - Adrar site n°03 (SD)	74,32	2,22	1,69E+15	0,05/1	**3,82**	0,93
Mimoun Adrar (SD)	72,00	1,21	1,10E+14	0,063/1,25	**3,60**	0,88
Cherouine -Timimoun (SD)	75,00	1,82	5,27E+14	0,05/5,00	**3,24**	0,98
Ouinna -Adrar (SC)	77,90	1,67	4,84E+12	0,063/6.3	**1,14**	0,80

Table 2.
Conventional and unconventional parameters for certain sands of the Adrar-Algeria sites [1].

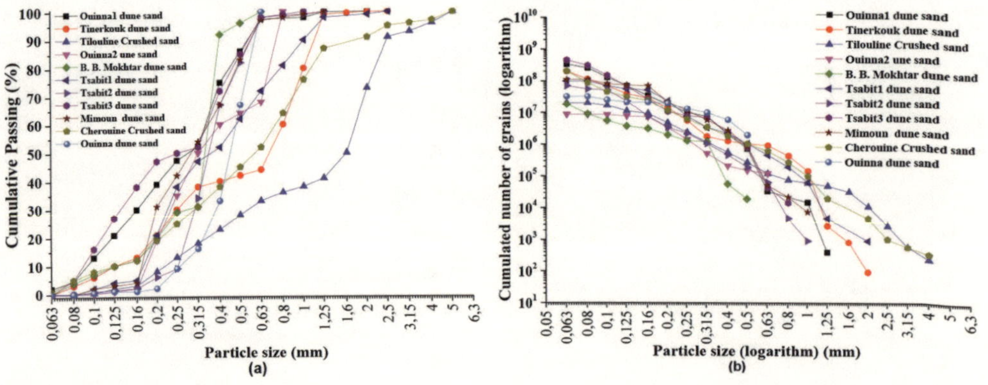

Figure 7.
Transformation of the particle size curves of Adrar Dune Sands to fractal lines a) particle size curves, b) fractal line.

Note: Physically, the value of the fractal dimension does not exceed the value of 3, because the study is within the framework of the three dimensions. But mathematically, that is to say according to the calculations of the hypothesis adopted, certain granular varieties can exceed the value of their fractal dimension the value 3.

But keep the optimal value that he adopted by chouicha [2], in his work, which does not exceed the value of 3 until the contrary is physically proven.

In **Figure 7 (a)** below, we show the grain size curves for all classes of dune sands as well as the quarry sand which we presented in **Table 2**, **Figure 7 (b)** we show the resulting fractal lines fractal distributions after transformation of their grain size curves for the different grain classes of dune sands. We note that the grain distribution of these sands is different and also gives us a different fractal distribution, which is proved by the obtained values of the fractal dimension.

Granular mixtures of concrete	**Cement class**		**Concrete mix**		**RC28**
	DFC	**R^2**	**DFb**	**R^2**	**MPa**
Mixture 01- [ARTS 22], CRS 42.5, Ouinna DS, Koussane Gr	2,85	0,98	2,66	0,99	22,80
Mixture 02- [ARTS 22], CPJ 42.5, Ouinna DS, Koussane Gr	2,75	0,99	2,83	0,99	20,60
Mixture 03- [LECT 24], CRS 42.5, Tinerkouk DS, Koussane Gr	2,94	0,98	2,61	0,98	17,90
Mixture 04- [LECT 24], CRS 42.5, Tillouline DS, Koussane Gr	2,94	0,98	2,76	0,99	18,90
Mixture 05- [LECT 24], CRS 42.5, Ouinna DS, Koussane Gr	2,94	0,98	2,85	0,99	18,40
Mixture 06- [LECT 24], CRS 42.5, DS Gr from B.B.M	2,94	0,98	2,87	0,99	19,10
Mixture 07- [LECT 24], CRS 42.5, Ouinna CS, Cherouine Gr	2,94	0,98	2,86	0,98	20,20
Mixture 08- [LECT 24], CRS 42.5, Brinkane DS, Cherouine Gr	2,94	0,98	2,83	0,99	29,50
Mixture 09- [LECT 24], CRS 42.5, Brinkane DS, Koussane Gr	2,94	0,98	2,76	0,99	20,50
Mixture 10- [LECT 24], CRS 42.5, Brinkane DS, Cherouine Gr	2,94	0,98	**2,79**	**0,99**	20,30
Mixture 11- [LECT 24], CRS 42.5, Brinkane DS, El Menia Gr	2,94	0,98	**2,79**	**0,99**	23,40
Mixture 12- [LAMCO 23], CPJ 42.5, Mimoun DS, Koussane Gr	2,75	0,99	**2,80**	**0,99**	18,80

***B.B.M:** Badji Badji Mokhtar. **El Menia,** Ghardaia. DS: Dune Sand Gr: Gravel.*

Table 3.
Identification by DF for granular mixtures of concrete at Adrar [1].

4.3 Particle size curves of mixtures

In this study, we adapted 12 concrete formulation tests for 12 projects carried out at the wilaya in the State of Adrar which are detailed in **Table 3**. Example of the particle sizes of the Granular Mixtures based on the particle sizes of different granular classes, **Figure 8** with cement and **Figure 8** without cement.

As for the concrete studies, we adopted the studies using dune sand that we have already studied, knowing that the concrete compositions included in this study are part of an executive study for projects scheduled to be implemented in arid regions.

Thus, the concrete compositions were studied in coordination between the building materials laboratory of the University of Adrar and the local technical laboratories, so that the study was carried out with the laboratory assigned to monitoring and control at the site. of the project. This within the framework of cooperation and the exchange of experiences between the university and the technical operator. These laboratories include the ARTS Soil Analysis Laboratory, Regional workshop of Saharan techniques, the LECT Technical Studies and Controls Laboratory and the LAMCO Building Materials analysis laboratory [26–28].

As for the aggregate materials used in these concrete structures, in addition to the sand dunes which were under study, the aggregates used locally come from quarries (Ouainna, Koussane, Cherouine, etc.), with two classes of CPJ cement. 42.5 and CRS 42.5.

In **Figure 8** we apply the hypothesis of fractal analysis and show through it an example of the granular mixture with its components, which is dedicated to the first concrete project, and through it we give a typical idea of granular mixtures applied in the field, giving us a clear picture of the conversion of the granular distribution into a fractal distribution from which we derive the fractal dimension of the concrete mixture, which is DF = 2.66 with a correlation coefficient of R^2 = 0.99. This granular mixture used, contains four granular classes, which are 03 classes of gravel and one of dune sand in addition to the cement class, which we have considered as a granular class in the dry concrete mix.

The variable component in our approved concretes is the variety of dune sands used, and we present in **Table 3** below a summary of the identification they concrete used through their dry granular mixtures used. The fractal dimension of concretes used in executive projects on site and not limited to laboratory research only. We draw attention to a necessary parameter, namely that we have taken into account the effect of cement as a granular material which participates in the

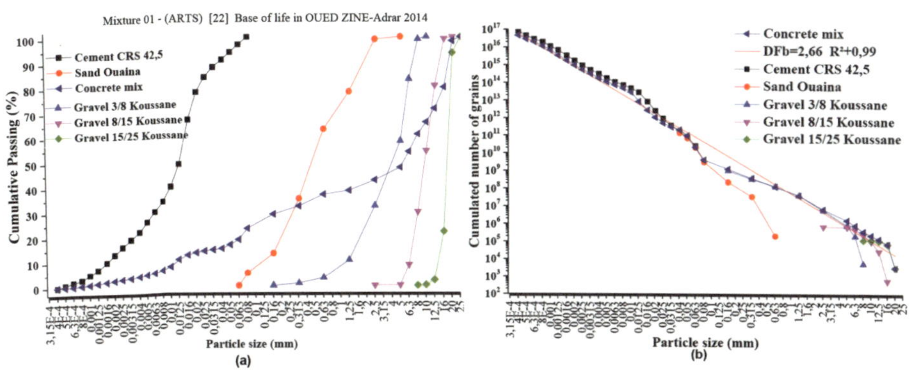

Figure 8.
Transformation of a grain size curve of the granular mixture of Oued zine to a fractal line (mixture 01) a) particle size curves, b) fractal line.

granular mixture by filling the pores with particles, while maintaining its main role in concrete, which is the main binder.

We have presented the values the fractal dimension of the cement used. Since we only used three classes of cement in the concrete compositions listed in **Table 2**, their values the fractal dimension of these classes, respectively, are as follows:

- CPj 42.5 DF = 2.75,
- CRS 42.5 DF = 2.85 Type 1,
- CRS 42.5 DF = 2.94 Type 1.

This is in accordance with what is indicated in the three cement classes mentioned in **Table 3**.

But it seems very clear that the difference in the values the fractal dimension of the granular concrete mixtures is due to all the different granular classes used in the concrete, which have shown their effect on the granular distribution in the granular mixture. The smallest value of the fractal dimension of concrete is DFb = 2.61 and the highest value obtained from the fractal dimension of concrete is DFb = 2.87. This confirms the previous results of researcher Chouicha [2], which indicates that the optimal value of the fractal dimension does not exceed the value 3.

5. Transformation of grain size curves using software

Table 1 shows details of how to transform a grain size curve into a fractal line for a single type of cement, and analytically we draw the fractal line of the cumulative grain sizes according to the grain dimensions with a scale logarithmic. But the granular mixture is made up of several different constituents, which makes it somewhat difficult to calculate the different arithmetic operations involved in fractal analysis.

The different required steps on which the proposed software is based can be addressed through two main sections.

First: enter the necessary data relating to each component of the mixture, namely the granular extent, the density and the standardized dimensions of the sieves, as well as the experimental data for the particle size analysis of each granular class.

Second: this software would draw the particle size curve of the granular mixture composed of several granular constituents, then display the transformed fractal line, and plot the linear fit and give its linear relationship followed by the correlation coefficient R2.

We proposed to name the program GranuFract in relation to the transformation of the particle size curve into a fractal line.

Figure 9 shows the image designed in the GranuFract software to capture basic information (density, cumulative sieve, and dimensions of sieve openings) to obtain the fractal distribution.

Concerning the transformation of the particle size curves to a fractal line, we adopted the method of fractal analysis carried out in the previous works [1, 29–31], as shown in **Figure 8**.

The GranuFract program allows us to easily determine the granular distribution of granular mixtures. This also helps us in the possibility of processing discontinu ous grain size curves, that is to say by knowing the coordinates of each grain. It also helps to infer the variance of the coordinates to correct for items that have missing items (**Figure 10**).

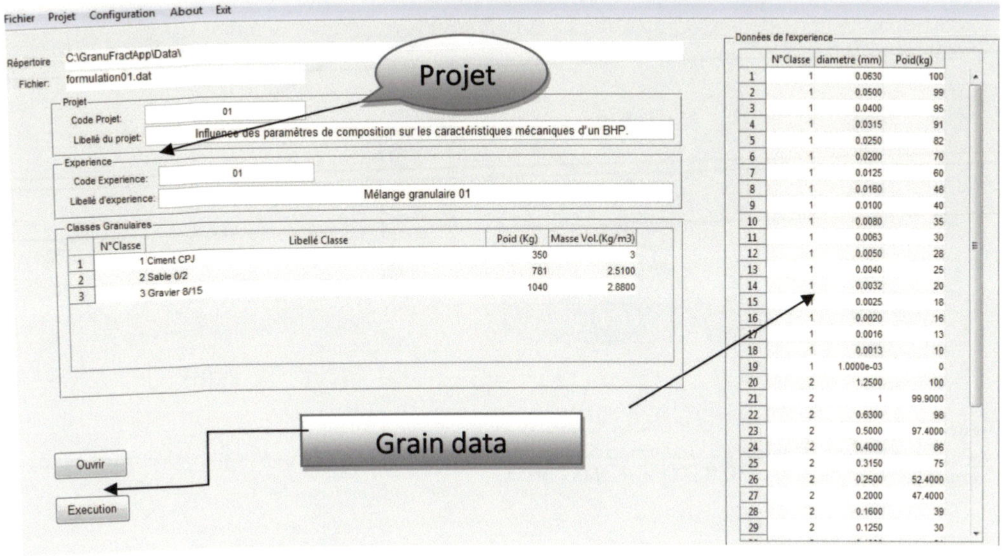

Figure 9.
Main window for entering basic information in the GranuFract software.

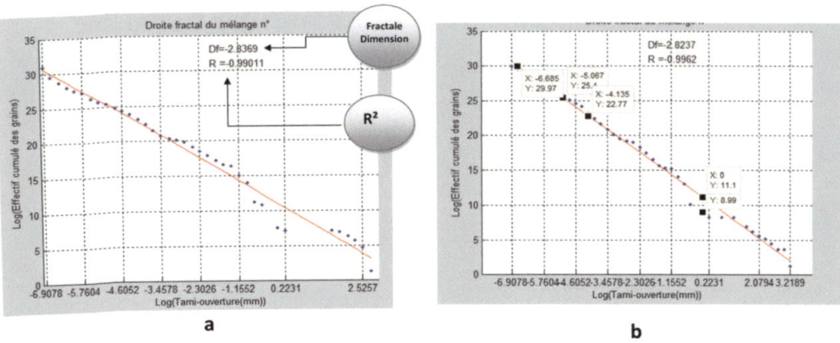

Figure 10.
Application of the transformation of a particle size distribution to a fractal distribution [1]. (a) Fractal line of the mixture. (b) Presentation of their coordinates.

6. Validation of software results

It is necessary to verify the data obtained from the program GranuFract, we have adopted ... the necessary procedures to compare the results that they obtained from the proposed software GranuFract and those that they obtained in the usual way, that is ie a graphical analysis using one of the mathematical programs "Origlab" and after a series of calculations The repeated values for each component of the granular mixture shown in **Figure 11 (a)** and **(b)** [1] we give the adopted values of the fractal dimension resulting from the two methods, they obtained values with an ideal correlation coefficient, and the standard deviation is ± 0.05. The values obtained by verification are:

According to 'GranuFract': DF = 2.82; R^2 = 0.99.

According to 'Originlab': DF = 2.83; R^2 = 0.99.

We draw the attention of those interested in such a study represented in the use of mathematical models to find solutions to the framework of the search for

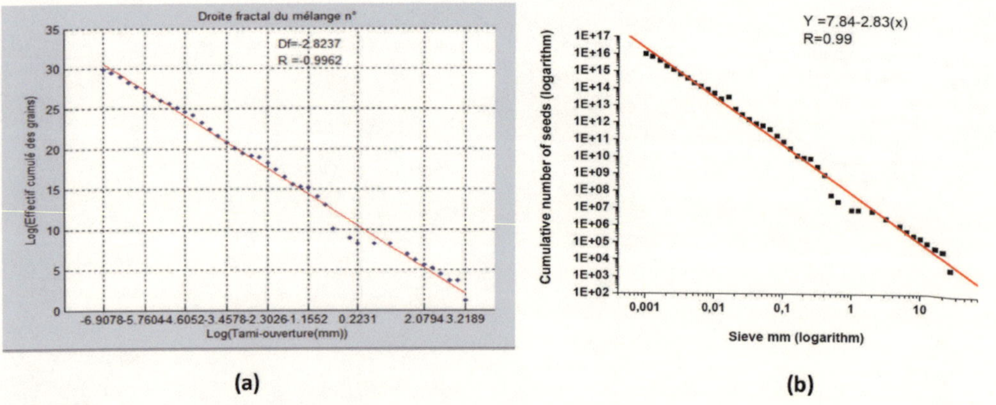

Figure 11.
Example of the validation of the transformation of a particle size curve to a fractal line [1]. a) Fractal line by GranuFract software, b) fractal line by originlab software.

concrete materials, that our main objective of this work is to highlight what we have achieved through to laboratory research and modeling in the application of the fractal analysis represented in the fractal dimension parameter and its positive results which have helped us to determine the granular distribution of the granular components of concrete through the granular mixture - the fractal line.

As for the proposed program, we look forward to improving and generalizing it by expanding a database that depends on a lot of experimental data for different components, including mineral additives and adjuvants, through which we hope to achieve a program to complete which will allow us to generalize it to all types of concrete, including innovative concrete.

7. Conclusion

This work allowed us to open a new window on the methods of concrete formulation using the granular distribution of granular mixtures to determine important properties in the definition of concrete, and we knew the efficiency of using modeling fractal to determine these granular mixtures of dry concrete for different concrete. It is now easy to know these granular mixtures according to their fractal dimension and their granular extent (FD, d/D).

Through this study, we show the contribution of an additional parameter for the formulation of concrete. It has already been mentioned that all concrete formulation methods depend on the granulation curve of the gravel mixture (sand gravel) to determine the gravel doses using the OAB grain reference curve, and the fractal model. Allow to determine the latter as a reference curve in the fractal dimension.

We initially proposed the "GranuFract" program, which converts the measurement curve of the grains of granular mixtures into a fractal line. This same program facilitates calculations and determination of the fractal distribution.

First of all, this work must be followed by careful empirical study to control the determination of granular aggregates with a concrete fractal line, in order to determine the dosages of the components of the concrete.

Secondly, we also aspire to extend this work by developing the "GranuFract" software in order to build an electronic database based on fractal modeling and obtained from particle size analysis in general.

Nomenclature

C	Cement dosage
E	Effective water dosage
d	Minimum grain size
D	Maximum grain size
d/D	Granular extent
E/C	Cement on water dosage report
G/S	Gravel on sand dosage report
N	Numbers of grains
N_c	Cumulative numbers of grains
EC	Cumulative workforce
ES	Sand equivalent
MF	Fineness modulus
ST_g	Total area of grains (mm^2)
FD	Fractal dimension
R^2	Correlation coefficient
SC	Crushed sand
SD	Dune sand
FDc	Fractal dimension of cement
FD_b	Fractal dimension of concrete mix
FD	Fractal dimension

Author details

Mhammed Abdeldjalil
Laboratory of Sustainable Development and Computer Science (LDDI), Civil Engineering, University of Ahmed Draia Adrar, Algeria

*Address all correspondence to: abdeldjalilmed@univ-adrar.edu.dz

References

[1] Abdeldjalil M, Yousfi S. dentification of sands of dune and concretes using a granular model - Case of arid region – Case Studies in Construction Materials 13 (2020) e00458. doi:10.1016/j.cscm.2020.e00458

[2] Chouicha K. La dimension fractale et l'étendue granulaire comme paramètres d'identification des mélanges granulaires, Mater Struct. 39 (2006) 665–681. doi:10.1617/s11527-006-9113-0

[3] De Larrard F. Structures granulaires et formulation des bétons. LCPC Nantes, France. 2000

[4] Gamil Y, and al, Simulation and Development of Instrumental Setup to Be Used for Cement Grouting of Sand Soil. Italian Journal of Science & Engineering Vol. 1, No. 12017.

[5] L. Qing, Q. Qingli, Z. Jun, W. Jiyang, Z. Qiang. Fractal dimension of concrete incorporating silica fume and its correlations to pore structure, strength and permeability Constr Build Mater 228 (20) (2019) 116986. https://doi.org/10.1016/j.conbuildmat.2019.116986

[6] L. Dan, N. Ditao, F. Qiang,L. Daming. Fractal characteristics of pore structure of hybrid Basalt–Polypropylene fibre-reinforced concrete. Cem Concr Compos, 109 (2020) 103555 https://doi.org/10.1016/j.cemconcomp.2020.103555

[7] Z. Di, S. Weidong, F. Jianxin, X. Gaili, L. Jiajian, C. Shuai. Research on mechanical characteristics, fractal dimension and internal structure of fiber reinforced concrete under uniaxial compression. Constr Build Mater 258 (20) (2020) 120351. https://doi.org/10.1016/j.conbuildmat.2020.120351

[8] Z. Peng, G. Zhen, S. Yan, L. Yuqiang, L. Jiazheng. Effect of large broken stone content on properties of roller compacted concrete based on fractal theory. Constr Build Mater (30) (2020) 120821 https://doi.org/10.1016/j.conbuildmat.2020.120821

[9] A. Rezaie , JP .Mauron Antoine, K. Beyer, Sensitivity analysis of fractal dimensions of crack maps on concrete and masonry walls, Autom. Constr. 117, (2020) 103258. https://doi.org/10.1016/j.autcon.2020.103258.

[10] Y. Xu ang, W. Mingzhi. Fractal dimension analysis of aggregate packing process: A numerical case study on concrete simulation. Constr Build Mater 270 (8) (2021) 121376 https://doi.org/10.1016/j.conbuildmat.2020.121376

[11] Féret R. Sur la compacité des mortiers hydrauliques. Annales des Ponts et Chaussées. Édition Paris, Vve C. Dunod, Série 7 vol. 4, 1892; pp. 5-164.

[12] Fuller. S.Thompson Proportion for concrete. Am Contractor, pp28-66,. 1907

[13] Abrams Duff A. Design of Concrete Mixtures. Bulletin No. 1, Structural Materials Research Laboratory, Lewis Institute, Chicago, 20 pp, 1918, (consulted le 2016). https://www.forgottenbooks.com/es/download/DesignofConcreteMixtures_10276916.pdf.

[14] Bolomey J. Granulation et prévision de la résistance probable des bétons, Bulletin technique de la Suisse romande, N°7, 62eAnnée, (1936) pp. 73-78. https://www.e-periodica.ch/cntmng?pid=bts-002:1936:62:95

[15] Caquot A. Role of inert materials in concrete. Memory of the society of civil engineers of France. 1937

[16] Faury J. Concretes: Influence of its inert constituents, rules to adopt for its best composition, its preparation and its

transport to construction sites: Hachettes. 1944

[17] Joisel A. Compositions of hydraulic concrete" Annals of the ITBTP (58), Series: Concrete and reinforced concrete. 1952

[18] Cassagnabere F. Produits prefabriques en béton file : vers l'amélioration des performances du matériau pour mieux gérer le procède de production. [thesis] Université de Toulouse III France. 2007

[19] Dreux G, Festa J. " Nouveau guide de béton " Eyrolles. 1995

[20] Baron Lessage J. The composition of hydraulic concrete: from the laboratory to the site. LCPC research report (64). 1976

[21] NF P18 452 "Measurement of the flow time of concretes and mortars with the maneuverimeter", AFNOR sagaweb,

[22] B. Mandelbort, Fractal objects, shape, chance and dimension. Flammarion Editions

[23] Sebsadji S.K, Chouicha K. Determining periodic representative volumes of concrete mixtures based on the fractal analysis, J Sold Struct, 49 (2012) 2941–2950.

[24] Marmi S. Dimension Fractal, [thesis] ; Mohamed Khider University, Biskra Algeria 2019

[25] Lecomte A, Thomas A. Caractère fractal des mélanges granulaires pour bétons de haute compacité. Mater Struct. 25 (5) (1992) 255–264. https://doi.org/10.1007/BF02472666

[26] Ansari A. ARTS, study report on the composition of the project concrete in Adrar, Algeria 2014

[27] Benhassen Y LECT, study report on the compositions of concrete at (Adrar, Tinerkouk, Tsabit, Cherouine, B B Mokhtar, Inzegmir), Adrar Algeria. 2012

[28] Mounir A. LAMCO, study report on the composition of the project concrete in Adrar, Algeria. 2013

[29] Abdeldjalil M. La compacité optimale des mélanges granulaires secs; application de l'analyse fractale. [thesis]; University of the U.S.T.O. Oran Algeria 1999

[30] Achouri F. Lecture des méthodes de formulation du béton à l'aide du modèle fractal de l'empilement granulaire, [thesis] ; University of the U.S.T.O. Oran Algeria 2011

[31] Bouregba A. Benarba M. Analyse fractale et compacité des mélanges granulaires secs, application au béton de sable. [thesis] ; University of the U.S.T. O. Oran Algeria 1998.